Monitoring Toxic Substances

Monitoring Toxic Substances

Dennis Schuetzle, EDITOR

Ford Motor Company

Based on a symposium sponsored

by the ACS Division of

Industrial and Engineering Chemistry

at the 174th Meeting of the

American Chemical Society,

Chicago, Illinois, August 31, 1977.

ACS SYMPOSIUM SERIES **94**

AMERICAN CHEMICAL SOCIETY

WASHINGTON, D. C.   1979

Library of Congress CIP Data

Monitoring toxic substances.
  (ACS symposium series; 94 ISSN 0097-6156)

  Includes bibliographies and index.

  1. Pollution—Measurement—Congresses. 2. Pollu-
tion—Toxicology—Congresses.
  I. Schuetzle, Dennis, 1942-   . II. American Chemi-
cal Society. Division of Industrial and Engineering
Chemistry. III. Series: American Chemical Society.
ACS symposium series; 94.

TD172.5.M66   615.9'07   78-27490
ISBN 0-8412-0480-2   ASCMC 8   94 1–289 1979

ACS Symposium Series

Robert F. Gould, *Editor*

FOREWORD

The ACS SYMPOSIUM SERIES was founded in 1974 to provide a medium for publishing symposia quickly in book form. The format of the Series parallels that of the continuing ADVANCES IN CHEMISTRY SERIES except that in order to save time the papers are not typeset but are reproduced as they are submitted by the authors in camera-ready form. Papers are reviewed under the supervision of the Editors with the assistance of the Series Advisory Board and are selected to maintain the integrity of the symposia; however, verbatim reproductions of previously published papers are not accepted. Both reviews and reports of research are acceptable since symposia may embrace both types of presentation.

CONTENTS

PREFACE

The recently legislated Toxic Substances Control Act (January 1, 1977) has emphasized the need for evaluating the effect of chemicals on the environment (risk assessment). A significant part of risk assessment consists of identifying the nature, concentration, and sources of chemicals released into the environment as a result of human activities as well as the biological effect of these chemicals.

These 16 chapters, which discuss many of the present state-of-the-art techniques for monitoring toxic substances in the environment, bring together numerous different approaches for identifying toxic substances. The papers are not meant to represent a comprehensive review of each particular area of monitoring but rather to describe recently developed methodology with several examples of applications.

Eleven of the chapters were presented at a symposium held at the 174th National ACS meeting in Chicago in 1977 chaired by the editor of this book and Ron A. Hites of the Massachusetts Institute of Technology. Five other chapters were added to increase the comprehensiveness of this volume.

The first two chapters describe the two most widely recognized techniques for the rapid biological screening of chemical substances for potential mutagenicity and carcinogenicity. The remaining chapters present instrumental analytical techniques for monitoring toxic substances.

In Chapter 1 Bruce N. Ames, the plenary speaker at the symposium, describes the development of the Salmonella/microsome assay, which is better known as the "Ames Test." This test was the first proven reliable and rapid in-vitro method for the mutagenic and carcinogenic screening of chemicals. Included is an excellent overview of the potential problem of human exposure to toxic substances as well as a description of how this important method is being used to test chemical substances.

New developments in cell-culture in-vitro assay tests for mutagenicity and carcinogenicity are presented by William G. Thilly in Chapter 2. It is expected that the cell-culture methodology will represent another important technique for the rapid biological assay of carcinogenicity and toxicity.

A complete environmental risk assessment, including extensive biological and analytical measurements, is not always required. Chapter 3 describes a phased approach developed for environmental assessments.

The philosophy of a preliminary survey analysis is used to help define the extent of potential risk (Level I analysis). Data gathered in the Level I survey analysis is used as a protocol for more detailed analysis (Level II). At Level III, a detailed analysis is made as a function of system parameters and time.

The next three chapters deal with the latest techniques for the analysis of organic pollutants in wastewaters and drinking waters. Instrumental analyses using liquid chromatography and gas chromatography/mass spectrometry (GC/MS) are described. Suitable sampling techniques for these pollutants are emphasized for subsequent instrumental analyses and bioassays.

Chapters 7 and 8 describe two major techniques for the monitoring of trace elements in environmental samples: atomic absorption (AA) and inductively coupled plasma–atomic emission spectroscopy (ICP). AA is most ideally suited for analyses where a limited number of trace metal concentrations are needed with high accuracy and precision. ICP has the advantage of multielement analysis with high speed.

The importance of surface analysis for evaluating the environmental effects of toxic substances is becoming more apparent as the result of recent work in this field. Chapter 9 describes ESCA, Auger, Ion Microprobe, and SIMS surface analysis techniques for atmospheric particulates. These techniques overcome the obvious limitations of bulk analysis, that is, the wide variability in the physicochemical characteristics of different particles.

Chapters 10 and 11 describe two relatively new techniques, FT/IR and opto-acoustic spectroscopy, for low molecular weight gaseous pollutants. Both of these techniques have the advantage of direct multicomponent analysis on a gas phase sample. Recent techniques are described in Chapters 12 and 13 for the sampling and analysis of trace organic substances in the atmosphere using GC/MS. Monitoring techniques for both gaseous and particulate air pollutants are described.

Chapter 14 describes the use of a new technique, ion chromatography, for the rapid measurement of ionic species in both air and water pollutants. It is expected that this method will replace a number of classical techniques currently used for the analysis of ionic species in environmental samples.

The need for specific detectors for the monitoring of toxic substances is becoming more important as potential hazards associated with particular toxic substances such as nitrosoamines are defined. The development of a specific detector for nitrosoamines and review of their formation and sources are described in Chapter 15.

Chapter 16 describes an information computer system which is being developed at the National Institute of Health and Environmental Protection Agency to bring past and future information on the monitoring of toxic substances into a highly accessible body of information. This computer system consists of a collection of chemical and biological data bases with a battery of computer programs for interactive searching. This system can be used to handle effectively the large body of data that will be generated during the next several years from the analytical and biological monitoring techniques described in this volume.

Ford Motor Company
Dearborn, Michigan
November 17, 1978

DENNIS SCHUETZLE

The Detection and Hazards of Environmental Carcinogens/Mutagens

BRUCE N. AMES'

Biochemistry Department, University of California, Berkeley, CA 94720

Since the 1950's we have been exposed to a flood of chemicals which were not tested for carcinogenicity or mutagenicity before use — from flame retardants in our children's pajamas to pesticides accumulating in our body fat. There are now about 35,000 commercial chemicals being made on a regular basis and 1,000 new ones are added every year. Even if only 1% of the known chemicals are mutagens or carcinogens, this could represent a serious problem. In the past, this problem was largely ignored, and even very large volume chemicals, involving extensive human exposure, have been produced for decades without adequate carcinogenicity or mutagenicity tests, e.g., vinyl chloride (2.5 billion kg/year, U.S.), ethylene dichloride (3.5 billion kg/year, U.S.), and a host of pesticides. A small fraction of these chemicals is now being tested in animals, but, for the vast majority of them, the only experimental animal is still the human, and epidemiologic studies on humans are impractical in most cases. As the 20- to 30-year lag time for chemical carcinogenesis in humans is almost over, a rapid increase in human cancer may be the outcome if too many of the thousands of new chemicals to which humans have been exposed turn out to be powerful mutagens and carcinogens. There has been much more concern in recent years about the problem of human exposure to man-made chemicals, though it is not an easy task to work out an adequate system for protecting the public.

There is increasing evidence that environmental chemicals, both man-made and natural, play an important role in the causation of cancer. In fact, the only known causes of human cancer aside from modifying genetic factors are radiation and chemicals, either from occupational sources, e.g., vinyl chloride, asbestos, chloroprene,[1] and coal tar,[2] from natural sources, e.g., aflatoxin and cigarette smoking, or from drug exposure.[1] In addition, many dietary factors have been linked to human cancer and these are likely to be chemicals, both natural and man-made.[1] If chemicals, both man-made and natural, are indeed the primary cause of human cancer, then pinpointing the most important offenders and minimizing human exposure to them may be the most effective means of reducing the incidence of cancer.

' Rather than have my tape-recorded talk from the ACS Symposium printed in this volume, portions of the talk, a paper by Dr. J. McCann and myself from the Origins of Human Cancer (Reference 6) and recent publications (References 18, 20, 24–26) have been combined to cover much of the same material in a more polished way. I am indebted to Dr. Dennis Schuetzle for the extensive editing.

Rapid Screening Test for Mutagenicity

We have been involved over the last 14 years in the development of a simple test[3] which can be used to identify chemical carcinogens and mutagens. The test detects these chemicals by means of their mutagenicity (ability to damage DNA, the genetic material) and is about 90% accurate in detecting carcinogens as mutagens.[4-6] Special strains of Salmonella bacteria are used for measuring DNA damage and these are combined with tissue homogenates from rodents (or humans) to provide mammalian metabolism of chemicals. The compounds to be tested, about 1 billion bacteria of a particular tester strain (several different histidine-requiring mutants are used), and rat (or human) liver homogenate are combined on a petri dish. After incubation at 37°C for 2 days, histidine revertants are scored.

Figures 1 and 2 show examples of the types of results obtained. In the "spot test" (Fig. 1), a small amount of the chemical to be tested is placed in the center of the dish, the chemical diffused out into the agar, and revertant colonies appear as a cloud around the central spot. This method is somewhat limited in sensitivity[3] but is extremely rapid. Normally, one tests individual dose levels of a chemical in the more sensitive plate test, and quantitative dose-response curves can be generated as shown in Figure 2. The initial portion of these curves is almost always linear, and most mutagens are detected at nanogram dose levels.

Validation of the Test

Ideally, one would like to validate the efficiency of such a test for use as a predictive tool for the detection of potential human carcinogens by testing a large number of human carcinogens. There is, however, very little data available on chemicals that cause cancer in humans, although almost all of the organic chemicals that are known or suspected human carcinogens are mutagenic in the test.[6] However, this is by no means sufficient to validate the test because the number of chemicals is small and also because the test must not only positively respond to carcinogens, it is also vital that it give a negative response with noncarcinogens. We have therefore validated the test using a large number (about 300) organic chemicals, of many chemicals classes, which have been tested in the conventional animal (usually rodent) carcinogenicity tests.[4,5] This procedure is not ideal since there is some degree of uncertainty surrounding the adequacy of certain of the animal tests themselves for assessment of carcinogenic risk to humans, although, with the exception of arsenic, chemicals known to cause cancer in humans also cause cancer in animals.[1] One particular difficulty in the interpretation of animal tests is especially apparent in the designation of a chemical as a noncarcinogen. Negatives are difficult to prove in the best of systems, and the statistical limitations of the animal tests make them particularly vulnerable to error in the designation of a chemical as a noncarcinogen. This problem is compounded by the great heterogenicity of animal cancer tests in the past, which differ enormously in degree of thoroughness, quality, and protocol. Out of this variety of tests, most of which are inadequate by current standards, where does one draw the boundary for defining a "non-carcinogen"? Clearly, some method is needed that would permit a more quantitative evaluation of negative cancer data. Some kind of completeness index might be useful which would permit expression of negative data as a "less-than" figure that would take into account limitations of the particular experimental system, such as duration of the experiment, number of animals used, and dose.

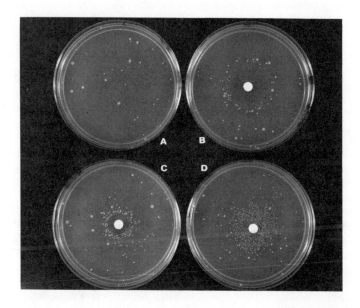

Mutation Research

Figure 1.   The "spot test." Each petri plate contains, in a thin overlay of top agar, the tester strain TA98 and, in the cases of plates C and D, a liver microsomal activation system (S-9 Mix). Mutagens were applied to 6-mm filter-paper discs which were then placed in the center of each plate: (A) spontaneous revertants; (B) furyl-furamide (AF-2) (1 μg); (C) aflotoxin B_1 (1 μg); (D) 2-aminofluorene (10 μg). Mutagen-induced revertants appear as a ring of colonies around each disc (3).

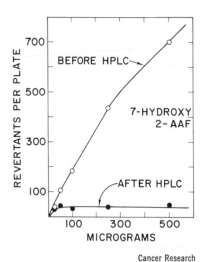

Cancer Research

Figure 2.   Mutagenicity of 7-OH-2-AAF before and after purification on HPLC. Each dose level was incorporated into top agar along with bacterial tester strain TA1538 and "S-9 Mix." Histidine revertants were scored after 48 hours incubation (3, 18).

Thus, in a validation such as ours, one expects some noncorrelation between mutagenicity and carcinogenicity due to the inadequacies of the animal cancer tests, especially in the case of "noncarcinogens" that show some mutagenic activity. In the interpretation of the results, therefore, it is crucial to determine what proportion of any noncorrelation is likely to be due to the animal tests and what proportion is due to the bacterial test. Even without such a determination, and in spite of the uncertainties due to the animal tests, the results show a striking correlation between carcinogenicity and mutagenicity. We found that 90% (157/175) of the carcinogens tested were mutagenic in the test, and 87% (94/108) of the "noncarcinogens" were non-mutagens.[4-6] In almost all cases, the test discriminated very well between carcinogens and noncarcinogens. This is especially striking with regard to a number of close chemical relatives where one is a potent carcinogen and the other is extremely weak or noncarcinogenic.[6] The relative mutagenicity of these chemicals in the test illustrates the ability of the test to discriminate efficiently between subtle differences in chemical structure which also drastically affect carcinogenicity. Our test system has also been independently validated with similar results, in a blind study by Imperial Chemical Industries[10] and in a study at the National Cancer Research Institute in Tokyo[11].

Among the chemicals which have been shown to be mutagens in Salmonella (as well as other short-term tests) that have recently been shown to be carcinogens are: 1,2-dichloroethane (10×10^9 lbs/year, U.S.), tris-dibromopropyl phosphate (the flame retardant used in children's polyester sleepwear), sulfallate (a pesticide), o-phenylenediamine, 2,4-diaminoanisole (hair dye ingredient), 2-nitro-p-phenylenediamine (hair dye ingredient), and 4-amino-2-nitro-phenol (hair dye ingredient).

We have not reported on any metal carcinogens, physical carcinogens such as asbestos, or radiation (although X-rays, fast neutrons, and ultraviolet light do mutate Salmonella bacteria). The standard test system is not suitable for metals entering the bacteria because of the large amount of Mg salts, citrate, sulfate, and phosphate in the minimal medium. A number of carcinogenic metals have been shown to be mutagens in bacteria using a somewhat different methodology from our standard procedure.[13, 14] At least some of the metals, such as chromate, mutate the Salmonella strains when salts in the medium are lowered (S. Rogers and G. Lofroth, pers. comm.).

We have previously discussed in detail the false positives and negatives in the test.[5,6] In general, we believe most of these are explainable by either inadequate animal carcinogenicity tests or inadequacies in our in vitro metabolic activation system. The limitations of both the animal cancer tests and the bacterial test in terms of maximum tolerated dose are defined by the toxicity of the chemical. In general, we have found that the expression of mutagenic activity occurs at concentrations well below the toxic level, and that toxicity is almost always associated with excessive DNA damage. However, the bacterial test could fail to detect chemicals that are very toxic to the bacteria for reasons not related to their mutagenic properties, for example, antibiotics. In fact, there were a small number (about 3%) of carcinogens and noncarcinogens that we could test only at very low dose levels because of their extreme toxicity.[5] This is also true of the animal tests, and, for this reason, some chemicals, such as TCDD and many of the chlorinated pesticides, can only be tested in the animal systems at extremely low doses.

The Salmonella test is currently detecting about 90% of organic carcinogens as mutagens. A number of important carcinogens are not detected however, particularly many heavily chlorinated chemicals.[5,6] We hope that with further improvements in both the tester strains and the metabolic activation system, as discussed above, the test might detect as much as 95% of all carcinogens.[5] The remaining few percent will never be detected by the Salmonella/microsome type of bioassay. We believe that most of these will fall into the hormone and promoter category or will be a class of mutagens unique to animal cells, such as griseofulvin, which specifically interacts with the mitotic apparatus.

There is a concentration of very weak mutagens among the "false positives." Of all the chemicals that have been positive in the test, about 25% have a mutagenic potency of less than 0.6 revertants/nmole, whereas of the "false positives", more than 60% are weak. We believe this suggests that most of the false positives may simply be weak carcinogens which have gone undetected in the animal tests. Recent work (15-17) indicates that there may be a rough correlation between carcinogenic and mutagenic potency. Thus a weak mutagen may actually be a weak carcinogen, and the statistical limitations of the animal tests make it more difficult to detect weak activity.

Millionfold Range for Mutagenic Potency

One of the most striking findings to come from testing a large number of chemicals in the Salmonella test is that we have found over a millionfold range of mutagenic potency of carcinogens in the test. This is illustrated in Table I.

TABLE I. Mutagenic Potency for Some Environmental Chemicals in the Salmonella/Microsome Test

	Revertants/nmole	Ratio
1,2-Epoxybutane	0.006	1
Benzyl chloride	0.02	3
1-Naphthylamine	0.42	70
Methyl methanesulfonate	0.63	105
2-Naphthylamine	8.5	1400
1-Aminopyrene	23	3800
2-Acetylaminofluorene	108	18,000
Benzo (a) pyrene	121	20,200
Aflatoxin B1	7,057	1,200,000
Furylfuramide (AF-2)	20,800	3,500,000

Date from Reference 4.

We observed significant numbers of mutagens throughout the entire potency range. If this range has any biologic reality in terms of mammals, it has some very important implications. One is the limits it sets on what constitutes significant variability in cancer biology. For example, it may not be highly significant if one chemical is a few times more or less carcinogenic than another, but it is important in terms of human risk if there is a 100-fold or a 1000-fold difference. We have also found a millionfold range in our analysis of carcinogenic potency (15,16). In terms of determining human risk, carcinogens with a millionfold difference in carcinogenic potency would certainly carry with them different human risks.

Mutagenic potency is dependent on the intrinsic mutagenic properties of the chemical, the efficiency of the *in vitro* microsomal activation system, and which tester strain is used (potency is calculated using results from the most active strain). Attempts are being made to determine what the relationship is between mutagenic potency and carcinogenic potency. Preliminary results indicate that, with a few exceptions, the *in vitro* test appears to be in rather good agreement (considering the million fold range in mutagenic and carcinogenic potency) with the animal tests. [15-17] Despite the complexity of the whole area of carcinogenic and mutagenic potency, we believe that it should be explored, though with caution.[16]

Mutagenic Impurities

We are finding that mutagenic impurities can play a significant role in mutagenesis testing. Because of the millionfold range (Table 1) in mutagenic potency observed in the Salmonella test even trace amounts of a mutagenic impurity in a non-mutagenic compound could be detected. As an extreme example, mg dose levels are required to detect the weakest mutagens and this defines the normal upper limit of the amount of a chemical tested in a standard test protocol. On the other hand, the most potent mutagens can be detected at a dose about 100,000 times less, in the ng range. Thus, in an extreme hypothetical case, a very potent mutagenic impurity, present at less than 0.001%, could conceivably cause a mutagenic response. And if this were present as a 1% impurity (not an unusual amount in most commercial grade chemicals) it could cause a mutagenic response when less than a μg of the non-mutagenic, unpurified chemical was tested. In our work, we have found a number of examples of mutagenic impurities. [18]

We have used high pressure liquid chromatography (HPLC) and the Salmonella/microsome mutagenicity test to look for mutagenic impurities in 8 aromatic amines and 3 polycyclic hydrocarbons [18](See Dorsey and Jones, this volume for a description of HPLC separations of toxic substances). We compared the mutagenicity of each of the 11 chemicals before and after purification on HPLC, and in 4 of the 11 cases examined we found a significant change in the mutagenic activity after purification. After purification of 7-hydroxy-2-acetyl aminofluorene almost all mutagenic ac tivity was lost (Figure 2). This original activity appears to be due to a 0.25% impurity of 2-acetyl aminofluorene.

In monitoring and evaluating the hazards of toxic substances, impurities are an area of concern. Industrial chemicals are likely to be a major source of mutagenic impurities. The Salmonella test can be used in the design of industrial syntheses and as a batch process monitor to minimize the introduction of mutagenic impurities.

Concentration and Assay of Mutagens

The Salmonella test can be used as a bioassay when separating and concentrating mutagens/carcinogens from complex mixtures such as feces, urine, cigarette smoke, and air and water pollutants. We have been developing techniques for concentrating the species of interest from urine using several different types of polymer resin.

We have been interested in analysis of mutagens in urine, [19-21] as have other groups. [22,23] The analysis of the mutagenicity of urine is complicated by several technical problems. Urinary metabolites are usually present in low concentrations and relatively little urine can be added directly to the Salmonella test system because urinary histidine interferes in the test. In addition, urine contains a variety of conjugates such as glucuronides and sulfates that are not split by the enzymes in Salmonella or in the mammalian liver added to the system. This latter problem can be overcome by adding glucuronidase (19,20,23) or sulfatase to the petri plate. In a recent study (20)we have explored the various types of adsorbent resins for concentrating mutagens from urine. Because most carcinogens and mutagens likely to be effective must penetrate the cell membrane, they tend to be nonpolar unless they are taken up by specific transport systems. Thus the use of a resin such as XAD-2 was particularly convenient because it had been previously shown to be quite effective at adsorbing relatively nonpolar compounds in water or urine, while letting histidine pass through. We added various mutagens/carcinogens to human urine and good recoveries have been measured after adsorption on XAD-2, XAD-4, and Tenax GC (diphenyl-p-phenylene oxide polymer). Even some of the conjugates of nonpolar compounds, such as glucuronides and sulfates, stick to the XAD resin.(20) Because one can concentrate the urinary materials several 100-fold, it is quite feasible to put the equivalent of 25 ml of human urine on a petri plate. The urine (up to 500 ml) is put through a column with a 1.5 cm^3 bed volume of XAD-2 (stryene-divinylbenzene polymer) and the adsorbed material is then eluted with a few milliliters of acetone. The acetone is taken to dryness and the residue is dissolved in dimethyl sulfoxide. This is the urine concentrate that is assayed for mutagenicity.

It has been shown with this procedure that cigarette smokers have mutagenic urine while nonsmokers do not. [20] One could use this method to examine the urine of a large population of nonsmokers for mutagenicity in an attempt to detect unsuspected mutagens/carcinogens. It would also be of interest to examine particular populations that are likely to be absorbing significant doses of mutagens, such as women dyeing their hair.[7] In cases where the mutagenic chemicals are known however, such as urine from children treated with add-on flame retardants [24-26] chemical methods are more sensitive. When new drugs are tried on humans this method should be used to examine the urine of the patients for mutagenicity as an adjunct to the standard toxicological tests in animals.

This method is of sufficient practicality to make it useful for routine screening of animals used in toxicology testing as an adjunct to the standard Salmonella plate test. There are several reports in literature of the detection of mutagens in animal urine that could not be detected in the standard plate test. [21,22]

The XAD-2 methodology is also useful in examining the mutagenicity of pollutants in water.[20,27] Several studies have shown the utility of XAD-2 for adsorbing nonpolar chemicals from water, referenced in 20. It also should be useful in examining a variety of other aqueous fluids that are consumed by humans.

Applications of the Salmonella Test

We believe that the test can play a central role in a long-term toxic substances monitoring program aimed at identifying and minimizing human exposure to environmental carcinogens and mutagens. It is a complement to traditional animal carcinogenicity tests (which take 2-3 years and cost about $150,000) as it can be used in a variety of ways not feasible with the animal tests.

1. Chemical companies can now afford to test routinely all new compounds at an early stage of development so that mutagens can be identified before there is a large economic interest in the compound.

2. The mutagenicity of a chemical may be due to a trace of impurity, and such knowledge could save a useful chemical [18].

3. Complex environmental mixtures with carcinogenic activity can be investigated using the test as a bioassay for identifying the mutagenic ingredients; e.g., cigarette smoke condensate, [8] ambient air pollutants [9] and water pollutants [see Chapter 6, this volume].

4. Human feces and urine can be examined to see if ingested or inhaled substances are giving rise to mutagens. [20,28]

5. The variety of substances that humans are exposed to, both pure chemicals and mixtures, are being assayed for mutagenicity by hundreds of laboratories, e.g., water supplies, (Chapter 6, this volume), soot from city air, hair dyes [7] and cosmetics, drugs, food additives, food, molds, toxins, pesticides, industrial chemicals, and fumigants.

6. The sensitivity of the Salmonella test may make it particularly useful for detecting chemicals that have weak carcinogenic activity and that would be difficult to identify in animal tests because of statistical limitations.

7. The test is being used for setting priorities in selecting chemicals for carcinogenesis bioassay in animals.

Other Short-Term Tests for Mutagenicity

Since the development of the *Salmonella* test and the demonstration that almost all carcinogens are mutagens, there has been a tremendous resurgence of interest in other short-term test systems for measuring mutagenicity. Many such systems have been developed and some, including the use of animal cells in tissue culture and cytogenetic damage in cells in tissue culture, are quite promising and have been validated with a reasonable number of chemicals. In addition, a number of the old systems, such as mutagenicity testing in *Drosophila*, have become much more sophisticated (the first mutagens known, such as X-rays and mustard gas, were first identified in *Drosophila* before they were known to be carcinogens). In addition, the development of several tissue-culture systems with animal cells, having as an end point the "transformation" of the cells to a tumor cell, is an important advance.

No one of these short-term tests, however, is completely ideal; for example, most tests using animal cells in culture require the addition of liver homogenate, just as our bacterial test does, because the animal cells useful for these tests are not capable of metabolizing all foreign chemicals to active mutagenic forms. Many short-term systems that are being validated seem to be effective in detecting known carcinogens. Because each system detects a few carcinogens that others do not, the idea of a battery of short-term tests is now favored.

In the case of a substance like Tris, or the food additive AF-2, the combination of an extremely widespread human exposure to the chemical and a positive result in a number of short-term tests should have been sufficient evidence to stop its use, considering that alternatives were available. Yet Tris and AF-2 were not removed from the market until the results from animal cancer tests indicated that they were carcinogens. It is becoming apparent that a positive result in many of these short-term test systems is meaningful, and that the systems may not only be a complement to animal cancer testing but may also provide much additional toxicological information as well. We can not and should not ban every mutagen and carcinogen as there are too many of them and many are quite useful, but we must treat them with respect. Mutagens deserve respect even apart from their carcinogenicity, though I think it will be unusual for mutagens not to be carcinogens as well.

Conclusions

Damage to DNA by environmental mutagens (natural as well as man made) may be the main cause of death and disability in advanced societies. We believe that this damage, accumulating during our lifetime, initiates most human cancer, as well as genetic defects, and is quite likely a major contributor to aging [29,30] and heart disease. [31,32] One solution is prevention: identifying environmental mutagens setting priorities, and minimizing human exposure. Rapid and accurate *in vitro* tests, such as the Salmonella/microsome test, should play an important role in realizing this goal.

References

1. Hiatt, H. H., Watson, J. D., and Winsten, J. A., Editors, "Origins of Human Cancer", Cold Spring Harbor Laboratory, Cold Spring Harbor, New York (1977).

2. Hammond, E. C., Selikoff, I. J., Lawther, P. L. and Seidman, H., "Inhalation of Benzpyrene and Cancer in Man", Ann. N. Y. Acad. Sci. (1976) *271*, 116.

3. Ames, B. N., McCann, J. and Yamasaki, E., "Methods for Detecting Carcinogens and Mutagens with the Salmonella/Mammalian Microsome Mutagenicity Test", Mutat. Res. (1975) *31*, 347.

4. McCann, J., Choi, E., Yamasaki, E., and Ames, B. N., "Detection of Carcinogens As Mutagens In the Salmonella/Microsome Test: Assay of 300 Chemicals", Proc. Natl. Acad. Sci. U.S.A. (1975) *72*, 5135.

5. McCann, J. and Ames, B. N., "Detection of Carcinogens As Mutagens In The Salmonella Microsome Test: Assay of 300 Chemicals: Discussion", Proc. Natl. Acad. Sci. U.S.A. (1976) *73*, 950.

6. McCann, J. and Ames, B. N., The Salmonella/Microsome Mutagenicity Test: Predictive Value for Animal Carcinogenicity. (p. 1431 in ref. 1)

7. Ames, B. N., Kammen, H. O. and Yamasaki, E., "Hair Dyes are Mutagenic: Identification of a Variety of Mutagenic Ingredients", Proc. Natl. Acad. Sci. U.S.A. (1975) *72*, 2423.

8. Kier, L. D., Yamasaki, E. and Ames, B. N., "Detection of Mutagenic Activity in Cigarette Smoke Condensates", Proc. Natl. Acad. Sci. U.S.A. (1974) *71*, 4159.

9. Talcott, R. and Wei, E., "Airborne Mutagens Bioassayed in Salmonella," J. Nat. Canc. Inst. (1977) *58*, 449.

10. Purchase, I. F. H., Longstaff, E., Ashby, J., Styles, J. A., Anderson, D, Lefevere, P. A., and Westwood, F. R., "Evaluation of Six Term Tests for Detecting Organic Chemical Carcinogens and Recommendations for Their Use", Nature (1976) *264*, 624.

11. Sugimura, T., Sato, S., Nagao, M., Yahagi, T., Natsushima, T., Seino, Y., Takeuchi, M. and Kowachi, T., "Overlapping of Carcinogens and Mutagens", In Fundamentals in Cancer Prevention (ed. Magee, P. N. et al.) University of Tokyo Press, Tokyo (1976).

12. Hartman, P. E., Hartman, Z., Stahl, R. C. and Ames, B. N., "Classification and Mapping of Spontaneous and Induced Mutations In the Histidine Operon of Salmonella", Adv. Gent. (1971) *16*, 1.

13. Venitt, S. and Levy, L. S., "Mutagenicity of Chromates in Bacteria and It's Relevance to Chromate Carcinogenesis", Nature (1974) *250*, 493.

14. Kreuger, C. S. and Hollocher, T. C., "Demonstration that divalent Be^{++}, Cu^{++}, Ni^{++}, and Mn^{++} are both point and frameshift mutagens in Salmonella," Mutat. Res. (1978) in press.

15. Sawyer, C. B., Hooper, K., Friedman, A., Peto, R., and Ames, B. N. (unpublished data).

16. Ames, B. N. and Hooper, K., "Does carcinogenic potency correlate with mutagenic potency in the Ames assay?" Nature (1978) *274*, 19.

17. Meselson, M. and Russell, K., "Comparisons of Carcinogenic and Mutagenic Potency," in Origins of Human Cancer (see ref. 1 for citation).

18. Donahue, E. V., McCann, J., and Ames, B. N. "Detection of Mutagenic Impurities in Carcinogens and Noncarcinogens by High-Pressure Liquid Chromatography and The Salmonella/Microsome Test", Cancer Research (1978) *38*, 431.

19. Durston, W. E. and Ames, B. N., "A Simple Method for the Detection of Mutagens in Urine: Studies with the Carcinogen 2-Acetylaminofluorene," Proc. Natl. Acad. Sci. U.S.A. (1974) *71*, 737.

20. Yamasaki, E. and Ames, B. N. "The Concentration of Mutagens from Urine with the Nonpolar Resin XAD-2: Cigarette Smokers have Mutagenic Urine", Proc. Natl. Acad. Sci. U.S.A. (1977) *74*, 3555.

21. McCann, J. and Ames, B. N., "Discussion Paper: The Detection of Mutagenic Metabolites of Carcinogens in Urine with the Salmonella/Microsome Test," Ann. N.Y. Acad. Sci. (1975) *269*, 21.

22. Legator, M. S., Connor, T. and Stoeckel, M. Ann. N. Y. Acad. Sci. (1975) *269*, 16.

23. Commoner, G., Vithayathil, A. J. and Henry, J. I. "Detection of metabolic carcinogen intermediates in urine of carcinogen-fed rats by means of bacterial mutagenesis," Nature (174) *249*, 850.

24. Blum, A. and Ames, B. N. "Flame Retardant Additives As Possible Cancer Hazards", Science (1977) *195*, 17.

25. Gold, M. D., Blum, A., and Ames, B. N. "Another Flame Retardant Tris-(1,3-Dichloro-2-Propyl) Phosphate and Its Expected Metabolites are Mutagens", Science (1978) *200*, 785.

26. Blum, A., Gold, M. D., Kenyon, C., Ames, B. N., Jones, F. R., Hett, E. A., Dougherty, R. C., Horning, E. C., Dzidic, I., Carroll, D. I., Stillwell, R. N., Thenot, J. P., "Tris-BP Flame Retardant is Absorbed by Children from Sleepwear: Urine Contains Its Mutagenic Metabolite, Dibromopropanol", (1978) Science, in press.

27. Hooper, N. K., Gold, C. and Ames, B. N. unpublished.

28. Bruce, W. R., Varghese, A. J., Furrer, R., and Land, P. C., "A Mutagen in the Feces of Normal Humans," (p. 1641 in ref. 1).

29. Burnet, F. M., "Intrinsic Mutagenesis: A Genetic Approach to Aging", Medical and Technical Publishing, Lancaster, England (1974).

30. Linn, S., Kairis, M. and Holliday, R. "Decreased Fidelity of DNA Polymerase Activity Isolated from Aging Human Fibroblasts", Proc. Natl, Acad. Sci. (1976) *73*, 2818.

31. Benditt, E. P. and Benditt, J. M., "Evidence for a Monoclonal Origin of Human Atherosclerotic Plaques", Proc. Natl. Acad. Sci. (1973) *70*, 1753.

32. Pearson, T. A., Wang, A., Solez, K. and Heptinstall, R. H., "Clonal Characteristics of Fibrous Plaques and Fatty Streaks from Human Aortas", Am. J. Palhol. (1975) *81*, 379.

RECEIVED November 17, 1978.

Human Lymphoblasts: Versatile Indicator Cells for Many Forms of Chemically Induced Genetic Damage

WILLIAM G. THILLY and JOHN G. DELUCA

Genetic Toxicology Group, Department of Nutrition and Food Science, Massachusetts Institute of Technology, Cambridge, MA 02139

Today I would like to discuss several different approaches to the determination of genetic hazards; I will try to draw attention to the relationship between the actual biochemistry of an animal and the probability that a particular chemical will cause genetic damage in that animal. Figure 1 summarizes a series of processes that determine whether or not a particular environmental chemical causes a particular unfortunate biological endpoint in the human population. This is a kind of paradigm, which we use to teach our graduate program in toxicology; a separate course is devoted to each process.

The first processes we consider are environmental production and distribution, and the probable modes of human exposure. Then we consider the distribution to tissues; even though this area of inquiry, pharmacokinetics, has not yet had a tremendous impact on deciding whether or not a chemical represents a public health hazard, models of physiological distribution are important in understanding toxicology in humans.

The next process we consider is drug metabolism. The study of oxidation reactions in compounds such as benzo (a) pyrene has emphasized the P450-dependent oxidation systems of the liver; toxicologists have virtually ignored other potentially important oxidizing systems. Future research may find drug-oxidation systems which are not P450-linked, and even other membrane-bound and non-membrane-bound oxidative and reductive systems in whole animal cells.

Chemical reaction with target molecules is another important consideration. It is true that *Salmonella typhimurium* uses the same four bases for DNA as do human cells; however, it is not true that the DNA is the same. The DNA inside bacteria is organized in a relatively simple fashion involving DNA-binding proteins, and, at some time in the growth cycle, adherence to the cell membrane. In human cells, however, the DNA is sequestered with proteins and coiled in a series of repetitive units, which are themselves arranged to make chromatin. Chromatin condenses at certain points in the growth of the human cell, forming the chromosomes. So, it does not seem unreasonable to expect differences in the chemical susceptibility and chemical reactivity of human and *S. typhimurium* genetic material.

Our next consideration is DNA repair. One tenet of the short-term testing business today is that there does not seem to be any reason to expect differences among mice, rats, and humans. In fact, there are important differences between humans and

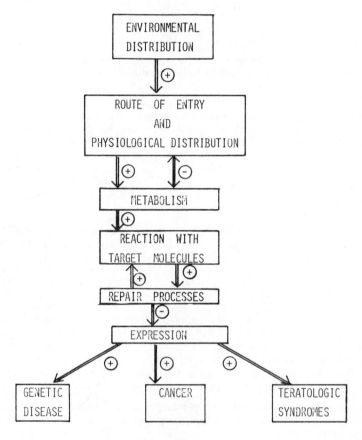

Figure 1.   A toxicological paradigm of environmental carcinogenesis and muta-genesis

rodents in the way they repair DNA. Rauth (1) demonstrated that if rodent cells are irradiated with ultraviolet light, and a little caffeine is then added, the observed toxicity indicates a marked synergistic effect. If the same thing is done with human cells, one sees the killing expected for the ultraviolet light, and an additional killing for the caffeine, but not the synergistic effect. It seems that caffeine is specifically inhibiting a major DNA repair mechanism (post-replication repair) in rodent cells, but this system is simply not present, or is certainly not a major dynamic feature, in normal human DNA repair (2-4).

Finally, we consider physiological expression or suppression, and the long latent period between carcinogen treatment and tumor appearance. While it makes sense to think in terms of phenotypic lag, clonal growth, vascularization, immuno-suppression, and other phenomena, we just do not know enough today to define clear-cut experimental approaches. However, at the human level, the undesirable endpoints of this series of processes are clear: teratologic syndromes, cancer, and heritable birth defects.

In summary, the reactions of people who, having been exposed to environmental chemicals, contract genetic disease or cancer are most complex. It is probably inappropriate to be satisfied with any of our present approximations; they are gross, and some are demonstrably incorrect. They are, however, "state-of-the-art," and constitute the only working models available to regulatory agencies and industry at the present time.

Now, I would like to go beyond the state of the art to raise a point which is often overlooked in the discussion of genetic disease and exposures that might be dangerous to humans. We will discuss four genetic endpoints [gene locus mutation, chromosome breaking (clastogenesis), recombination, and DNA repair], considered here in terms of assaying for genetic hazards.

First, let us differentiate between gene locus mutation, a specific kind of genetic damage, and the general question of genetic damage in humans. A gene locus mutation is a localized change in the sequence of DNA, e.g., an adenine is replaced by a guanine; an AAA sequence is replaced by an AA sequence by deleting a base pair. There are many chemicals which cause gene locus mutation, and they are specifically called mutagens; they represent one class of genetic hazards. But in human cells, as in those of all higher animals, DNA is organized into chromosomes. Most frequently, breaking a chromosome kills the cell, but not always, and a cell which survives a chromosome break is genetically different from its ancestors.

Another kind of genetic endpoint presently under consideration for regulation is recombination. Recombination in a molecular sense is a reciprocal exchange of DNA between homologous chromatids, and is detected by changes in the association of genetically linked markers.

The fourth endpoint is DNA repair. One can determine whether DNA has been damaged in a chemical sense by measuring, by any of several techniques, whether certain characteristic changes in DNA, mainly strand breakage or resynthesis, take place after treatment.

Having briefly defined these four endpoints, I would like to ask, "Are they related to known human disease?"; "Can they be measured quantitatively?"; and "Can they be measured with facility for inclusion in an overall plan for environmental bioassay?" Table I lists examples of gene locus mutations resulting in human disease: phenylketonuria, histidinemia, Hartnup disease, and cystinuria. Phenylketonuria is detected by the presence of phenyl ketones in infant urine, and we are able to overcome this genetic defect by appropriate dietary measures in early life. Unfortunately, this is not the case with many other genetic diseases. Listed are just a few dealing with amino acid metabolism. There are now about two hundred amino acid metabolism diseases identified in humans, and a sizable portion are associated with diminished mental capacity. These are very unfortunate diseases to have in the family, and they are a very expensive social problem. Note that there is a frequency of occurrence of about one in 10,000 to one in 30,000 live births. Since there are actually a thousand or so human diseases of putative gene locus mutation origin, a first-order approximation would indicate that nearly 1% of surviving infants carry a clinically recognizable genetic error.

TABLE I

Examples of Human Gene Locus Mutation
(from Vogel [5])

Disease	Estimated Frequency
Phenylketonuria	1 in 11,500
Histidinemia	1 in 24,000
Hartnup Disease	1 in 26,000
Cystinuria	1 in 7,000

Figure 2 shows something interesting that I had not realized until I reviewed the work of Vogel (5), a German mutational epidemiologist, who examined the frequencies of human diseases that are strictly gene-locus in origin. He found that, as paternal age increased, there was an increase in the incidence of gene locus mutations. Most of us are acquainted with the correlation between Down's syndrome incidence and maternal age, but how many of us have ever noticed Vogel's equally compelling observation regarding gene locus mutation in humans? We may ask, "What is responsible for this paternal age dependence?" A trivial hypothesis arises from consideration of the human male reproductive system, in which spermatogonia continuously divide. Is it unreasonable to assume that, as a result of this kind of continuous division, the gene locus mutations would accumulate in the germ cells — the primal cells which continuously divide? One daughter cell becomes a sperm, and one remains near the vascular bed of the testes to continue to divide and create more sperm over the life of the male. The observed increase with paternal age could arise as a result of spontaneous aging, or of some kind of environmental exposure; there is absolutely no data to differentiate between these possibilities.

Figure 3 deals with gene locus mutation in S. typhimurium as an example of the kind of data one may expect. Bruce Ames's strains TA1535, TA1538, TA98 and TA100 were used separately to determine the mutagenic potency of aflatoxin B_1 in the presence of the postmitochondrial supernatant and appropriate cofactors. As an aside here, I might mention that I do not completely agree with Dr. Ames's original thesis that the use of a set of specific reversion assays is the best way to detect gene locus mutagens. I think we should consider that it might be better to use a relatively big genetic target containing many different types of mutable sites. Tom Skopek, a graduate student in my lab, was worried about this problem, so he developed a forward mutation assay in an attempt to overcome the necessity of dealing with five different

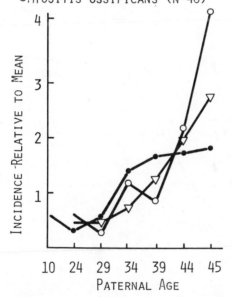

GENE LOCUS MUTATION & PATERNAL AGE
▽ACHONDROPLASIA (N=175)
●HEMOPHILIA A, MATERNAL
              GRANDFATHERS (N=77)
○MYOSITIS OSSIFICANS (N=40)

INCIDENCE -RELATIVE TO MEAN

PATERNAL AGE

*Figure 2.   Correlation of paternal age with incidence of human diseases caused
by gene locus mutation (redrawn from Ref. 5)*

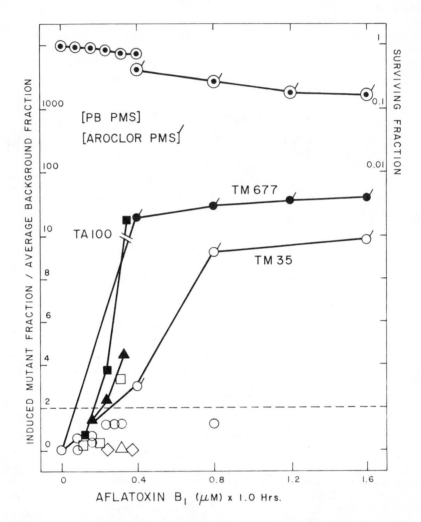

Figure 3. Reversion to histidine protrophy in Ames's S. typhimurium *strains (TA1535, TA1537, TA1538, TA98, TA100) and forward mutation to 8-azaguanine resistance in* S. typhimurium *strains TM35 and TM677 induced by aflatoxin in the presence of rat postmitochondrial supernatant. (□) TA1535; (△) TA1538; (◇) TA1537; (▲) TA98; (■) TA100; (○) TM35; (●) TM677; (⊙) Toxicity (all strains). Unflagged data represents phenobarbital-induced rat liver postmitochondrial supernatant (PB PMS).*

strains by use of a single strain assay. At this point, I think he has been successful (6). In Figure 3, it can be seen that Skopek's strain TM677 in an assay for mutation to 8-azaguanine resistance is indeed equally as sensitive as Ames's strain TA100 to afla-toxin. This new assay is also as senstive as the most sensitive reversion assay for arylating, alkylating and intercalating agents that we have examined so far. This bit of haggling over procedural detail aside, a gene locus mutation assay is essentially count-ing colonies on plates and determining whether treatment has caused a statistically significant increase in mutant fraction.

Figure 4 shows that gene locus mutation assays can also be performed using diploid human cells. Actually, this is what I do for a living — sit around and subject unfortunate human cells to mutagens. We have developed a system using diploid lymphoblast lines and mutation to 6-thioguanine resistance (7). In Figure 4 we see aflatoxin B_1 mutagenicity. Note by comparison to Figure 3 that the sensitivities of the human and bacterial cell mutation assays are quite similar.

The second type of genetic change we are considering is chromosomal abnormal-ity. Table II lists some examples with which you are already familiar, i.e., Down's syndrome and Kleinfelter's syndrome.

TABLE II

Examples of Human Disease Associated with Chromosomal Abnormality (from Vogel [5])

Disease	Chromosome Abnormality	Frequency
Kleinfelter's syndrome	XXY	1 in 400
Down's syndrome	trisomy 21	1 in 300 to 1 in 600
Cri du chat syndrome	deletion in chromo-some 5 with other structural rearrangements	1 in 200

These very unfortunate syndromes do not occur at frequencies of one in 10,000 or one in 30,000; they are occurring on the order of one in 500 to one in, say, 2,000. Again, if you add together the several hundreds of human syndromes that are apparently caused by chromosome aberration, you discover a very expensive public health prob-lem. I cannot offer any references to good estimates, but it seems that the total misery inflicted on the human population by chromosome aberration is approximately equal to that inflicted by gene locus mutation.

In Figure 5, I have again replotted some of Vogel's data (5) to demonstrate that maternal age correlates positively with the relative incidence of chromosome abnor-mality diseases such as Patel's syndrome or Edward's syndrome. Paternal age does not affect the probability of chromosome aberration in offspring. If I had plotted maternal age versus the frequency of gene locus mutations in humans, it would have been flat. A not unreasonable explanation might be that the different physiology of germ cell development in humans — males and females — leads to this different kind of expres-sion of gene locus mutation and chromosome aberration as a function of parental age.

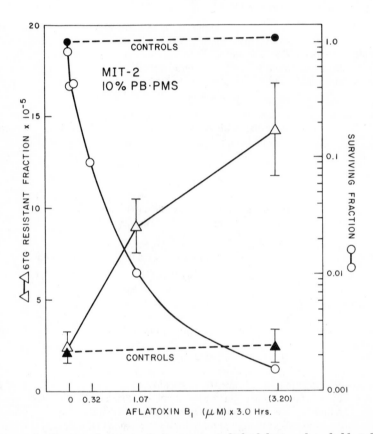

Figure 4. Toxic and mutagenic response of diploid human lymphoblast line MIT-2 to aflatoxin B₁. Open symbols: aflatoxin + phenobarbital-induced rat liver postmitochondrial supernatant (PMS); closed symbols: effect of treatment with either aflatoxin or PMS alone.

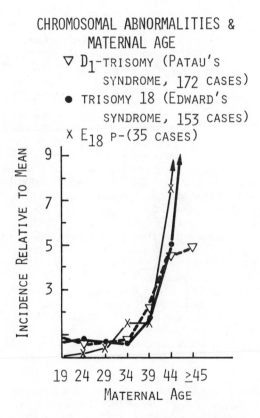

Figure 5.   Correlation of maternal age with incidence of human diseases caused by chromosome abnormalities (redrawn from Ref. 5)

There is a practical point to be considered here, since there are safety regulations which prevent women of childbearing age from exposure to suspected chemicals, and these regulations seem to be based on the premise that paternal exposure would not be expressed for some reason. This irrational regulatory status may be based on the belief that, since human sperm develop to maturity in about two weeks, exposed sperm will somehow disappear or possibly be killed. However, this approach ignores the fact that the basal germ cell stays there, and "remembers" all of its unrepaired, non-lethal chemical insults.

Figure 6 is an example of a quantitative study of chromosome breaks induced by methylnitronitrosoguanidine (8), which is also a potent gene locus mutagen. However, it is a good idea to investigate chemicals that are not both chromosome breakers and gene locus mutagens, but are only one or the other; this is just one of the difficult problems of setting up an appropriate set of assays to test whether or not something is a potential public health hazard.

Figure 7 takes on the question of recombination. I know of no human disease that is suspected of arising from reciprocal recombination. In mammalian cells, there is no direct way of measuring genetic recombination at the moment; there are very elegant techniques for use in *Neurospora*, in bacteria, and in other forms, but the difficulties of mammalian cell culture are still preventing the development of appropriate assays. However, there is a technique development of appropriate assays. However, there is a technique developed by Sam Latt (9) at the Children's Hospital in Boston, which is quite elegant. I thought I would mention it in case some of you were unacquainted with it. Basically, the cells will grow in the presence of bromodeoxyuridine (BUdR). If cells grow long enough in BUdR, all of the chromosomes will incorporate some BUdR, which will quench the fluorescence of DNA/dye complexes; this can be observed by fluorescence microscopy. One treats cells with the test chemical while incubating them with BUdR, and simply scores reciprocal exchange between chromatids on the same metaphase chromosome. Occasionally, several recombination events can be observed between the chromatids on the same chromosome. Many mutagens are both chromosome breakers and recombinogens. However, there is at least one example of a chemical that is reported to be a recombinogen in yeast, but is not a clastogen or a nuclear gene locus mutagen: fluorodeoxyuridine. Figure 8 is the plot of a concentration-dependence study of sister-chromatid exchange induced by phosphamid (10), which, given an appropriate drug-metabolizing element, is also a gene locus mutagen and a clastogen.

There are a number of different ways of using DNA repair as an endpoint; one example is the measurement of unscheduled DNA synthesis. One way in which human cells respond to chemical damage of DNA is by enzymatically excising a portion of DNA, including the chemical lesion, and filling in the resulting gap by polymerase and ligase action. Radioactive thymidine can then be used to detect repair synthesis by counting grains over lightly labeled (non-S phase) cells on an autoradiogram. DNA repair has many other facets that cannot be adequately treated here. Let me just refer in passing to the work of Louise Prakash at Rochester, who selected about 40 methyl methane-sulfonate-sensitive clones of *Saccharomyces*, and found, through complementation analyses, more than 30 independent complementation groups (11). In short, there seems to be a very large number of DNA-repair systems in *Saccharomyces*; this may reasonably be expected to be true for human cells. The principle regulatory premise seems to be that induction of metabolic activity involving the DNA is *prima facie* evidence of potential genetic damage.

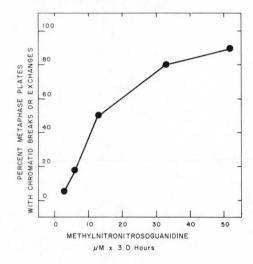

Figure 6. *Chromosome aberrations induced in human fibroblasts by nitrosoguanidine (MNNG) (redrawn from Ref. 8)*

RECOMBINATION

(a) Recombination has no clear relation to known
      human diseases.

(b) Genetic systems for direct measurement are
      still under development.

(c) Sister chromatid exchange is a technique
      which reveals exchanges between chromatids
      by differential staining.

(d) Many mutagens and clastogens are recombinogens.

Figure 7. *Recombination as an assay endpoint*

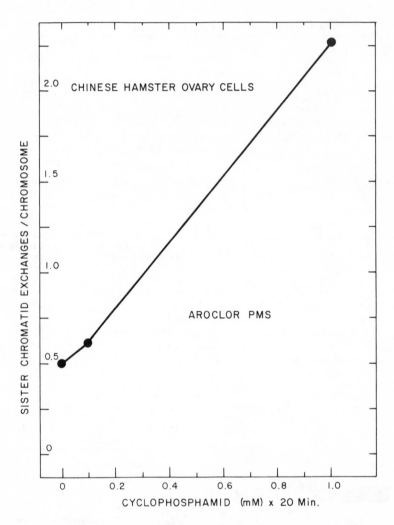

*Figure 8. Cyclophosphamid-induced sister chromatid exchange in CHO cells
(redrawn from Ref. 10)*

Figure 9 summarizes an example of a DNA-repair study with primary rat fibroblasts *(11)*. Unscheduled DNA synthesis has been measured using autoradiography. Cells were labeled with radioactive thymidine after exposure to the test agent, covered with a photographic emulsion, and allowed to stand several weeks before the film was developed. The number of grains appearing over the nuclei of these cells increases with the concentration and nature of the test agent. A grain indicates the presence of tritiated thymidine in the acid-insoluble nuclear material (DNA).

In concluding this brief review, I feel it is necessary to restate that it is not true that a recombinogen is a clastogen is a mutagen, and, I have no problem in adding, is a carcinogen. Obviously, I do not agree with the premise that one should look for one kind of genetic damage exclusively. The different genetic endpoints do seem to have different molecular etiologies, although we do not really know what they are. To measure any of these genetic endpoints, we have a wide choice of biological species: hamster, mouse, or rat cells, *E. coli, S. typhimurium, Tradescantia,* and a host of others. On occasion, however, I and other members of our laboratory have seen fit to argue, on the basis of biochemical similarity between human cells, that it would make sense to use human cells for all of these endpoints.

Figure 10 summarizes our suggestion of how a testing laboratory of appropriate competence could carry this out in a reasonably inexpensive way. The diploid human lymphoblast, because of its similarity to cells in the human body, can be used effectively in gene locus mutation assay, in the measurement of recombination (sister-chromatid exchange), in enumerating chromosome aberrations, in measuring DNA repair by unscheduled DNA synthesis or by other techniques, and of course in the direct measurement of cellular toxicity. The total elapsed time for a laboratory to obtain the data required in all of these assays would be about a month. Some of these assays take more worker time than others; for instance, the present technique for gene locus mutation assays requires seven full worker days. However, it takes only one worker day to measure toxicity, two to measure DNA repair, and so forth. I press this suggestion in the belief that there are remarkable differences in chemical/genetic susceptibility among rodent cells, human cells, and, incidentally, *Salmonella* cells.

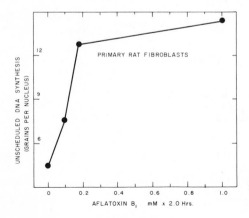

Figure 9.   Aflatoxin-induced UDS in primary rat fibroblasts (redrawn from Ref. 12)

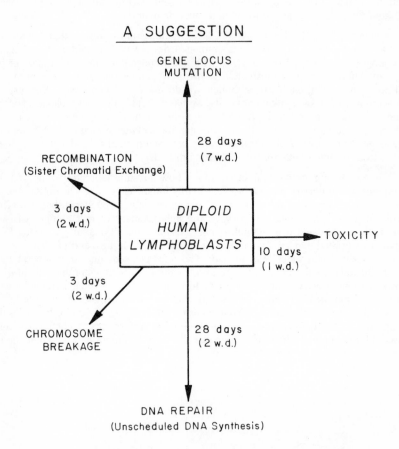

Total elapsed time 28 days (14 w.d.)

Figure 10. A scheme for regulatory testing using human cells and multiple end points

Literature Cited

1. Rauth, A. M., M. Tammemagi, G. Hunter, Biophys. J. (1974) *14*, 209.

2. Fujiwara, Y., T. Kondo, Biochem. Biophys. Res. Commun. (1972) *47*, 557.

3. Trosko, J. E., E. H. Y. Chu, Mut. Res. (1973) *12*, 337.

4. Buhl, S. N., J. D. Regan, Biophys. J. (1974) *14*, 519

5. Vogel, F., in "Chemical Mutagenesis in Mammals and Man" (F. Vogel, G. Roehrborn, eds.), p. 16, Springer Verlag, New York, 1970.

6. Skopek, T. R., H. L. Liber, J. J. Krolewski, W. G. Thilly, Proc. Natl. Acad. Sci. U.S.A. (1978) *75*, 410.

7. Thilly, W. G., J. G. DeLuca, H. Hoppe IV, B. W. Penman, Chem.-Biol. Interact. (1976) *15*, 33.

8. Stich, H. F., R. H. C. San, Proc. Soc. Exp. Biol. Med. (1973) *142*, 155.

9. Latt, S. A., J. Hist. Cytol. (1976) *24*, 24.

10. Stetka, D. G., S. Wolf, Mut. Res. (1976) *41*, 346.

11. Prakash, L., Mut. Res. (1976) *41*, 241.

12. Stich, H. F., R. H. C. San, P. P. S. Lam, D. J. Koropatnick, L. W. Lo, B. A. Laishes, I.A.R.C. Sci. Pub. (1976) *12*, 427.

RECEIVED November 17, 1978.

A Phased Approach for Characterization of Multimedia Discharges from Processes

J. A. DORSEY, L. D. JOHNSON, and R. G. MERRILL

Industrial Environmental Research Laboratory, Environmental Protection Agency, Research Triangle Park, NC 27711

In recent years, concern for the environmental impact of discharges from industrial and energy processes to air, water, and land receptors has expanded far beyond the traditional lists of a few selected pollutants. This has given rise to the concept of a comprehensive characterization of all materials released to the environment, and the phrase "environmental assessment" has been added to the vocabulary of many organizations. The four major components of an environmental assessment, as used by our group, are as follows: (1)

1. A systematic evaluation of the physical, chemical, and biological characteristics of all input and output streams associated with a process;

2. Predictions of the probable effects of those streams on the environment;

3. Prioritization of those streams relative to their individual hazard potential; and

4. Identification of any necessary pollution control technology programs.

This process is depicted schematically in Figure 1.

It is obvious that such an assessment is formidable, technically difficult, and extremely expensive. Since an environmental source assessment study is required to characterize the total pollution potential of all waste streams, the sampling program must be more extensive than those conducted for the acquisition of process or control engineering data. Assessment sampling is more complete in that all waste streams are sampled and no attempt is made to limit sampling to a preselected number of process streams. The sampling is also more comprehensive in that all substances of potential environmental concern must be detectable above some minimum level of concern. These requiements of completeness and comprehensiveness call for a strategy of approach in which the philosophy and structure ensure maximum utilization of available resources.

Approach

Two clearly distinct strategies to an environmental assessment sampling and analysis program that satisfy the requirements for comprehensive information are the direct and phased approaches. In a direct approach, all streams would be carefully sampled and the samples subjected to complete, detailed analysis using compound specific analytical techniques. In a phased approach, all streams would first be sur-

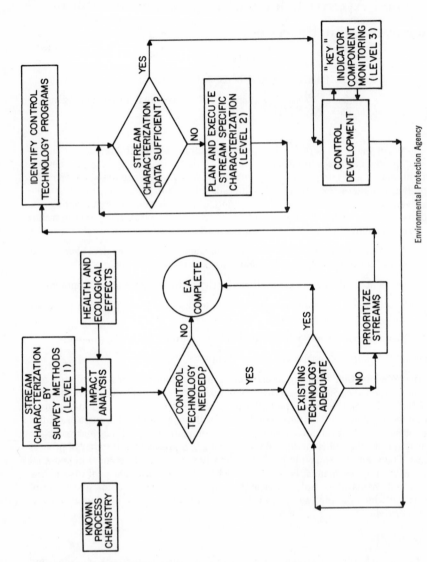

Figure 1. Flowchart of an environmental assessment process (1)

Environmental Protection Agency

veyed using simplified, generalized sampling and analytical methods which would permit the streams to be ranked on a priority basis; i.e., very hazardous streams would be distinguished from those less hazardous or relatively innocuous. Detailed sampling and analysis would then be applied first to streams ranked in the highest priority by the survey, and then to other streams in descending order of potential hazard. Another phase, initiated after complete characterization of potential problems, would involve continuous monitoring of "key" indicator materials to evaluate long-term process variability.

It should be clear that any partially direct approach (e.g., the use of predetermined lists of compounds) is not consistent with the complete and comprehensive requirements of either the direct or the phased environmental assessment philosophy and therefore is not an alternative to either approach. Similarly, *a priori* judgments based on process chemistry, thermodynamics, etc. are not acceptable practices in this context.

Since both the direct and phased approaches offer, at least in principle, equal promise for ultimate success (i.e., comprehensiveness and complete characterization), it is worthwhile to examine their relative resource requirements. Several studies were conducted with the objective of comparing the costs of direct and phased (with elimination of low priority streams) sampling and analysis approaches. (2,3) A number of processes were evaluated during these studies and the results for two unit operations — a limestone wet scrubber and full-scale low-Btu coal gasifier — are taken as examples. The scrubber involved seven feed or waste stream sampling sites. The gasifier contained 70 identifiable stream sampling points. The total estimated costs for both processes by both approaches are shown in Table I.

Table I

Estimated Total Costs of Direct vs Phased Approaches to Sampling and Analysis

Process	Phased	Direct
Limestone Wet Scrubber	$350,000	$ 500,000
Coal Gasifier	$725,000	$1,450,000

In both bases, the phased approach was found to be more cost effective than the direct approach even though the scrubber and coal gasifier differ markedly in size, complexity, basic technology, and total cost of sampling and analysis. The cost advantages of the phased sampling and analysis approach were found to be approximately proportional to the complexity of the process being sampled. Within the phased approach, the initial sampling and analysis costs were shown to be 10 percent of the total cost of the phased effort. Thus, many qualitative judgments, including whether or not a full-scale endeavor is at all necessary, can be made at low cost before a commitment is made to initiate a detailed assessment.

This actual resource savings is only one aspect of justification for the phased scheme. As the result of information developed in the survey phase, a significant improvement in sampling accuracy and completeness can be anticipated during later phases. It is very doubtful that equal data could actually be acquired in a direct approach due to the lack of the very necessary learning processes involved in difficult source sampling and analytical projects.

As the result of these studies, the phased approach was selected for further development and implementation. The strategy makes use of the three levels of sampling and analysis mentioned earlier. The survey phase is defined as Level 1; the detailed analysis of priority components, Level 2; and the monitoring of selected components, Level 3.

Sampling Programs in a Phased Approach

It has been noted that it is not sound practice to attempt to define a detailed sampling program until: (1) the general characteristics of the streams in question have been evaluated, and (2) the nature of any unfavorable sampling system/sample interactions has been considered (e.g., chemical reaction, volatility loss). Hence, an effective sampling program involves a series of reiterative tests in which each iteration enhances the source assessment by focusing resources and efforts on the pollutants and streams of concern and improves the accuracy of the sampling program.

Level 1 sampling stresses the concept of completeness by presuming that all streams leaving the process will be sampled. Level 1 sampling is not predicated on *a priori* judgments as to stream composition. The techniques utilized presume that whatever knowledge is available is, at best, incomplete. Predictive and extrapolative techniques employed during source assessments serve as a check on the empirical data and not as a replacement for them. Level 1 sampling systems are therefore envisioned to permit collection of all substances in the stream at a reasonably high level of efficiency. They do not necessarily produce information as to specific substances or their chemical form. Further, Level 1 sampling programs are designed to make maximum use of existing stream access sites. While some care must be exercised to ensure that the samples are not strongly biased, the commonly applied concepts of multiple point, isokinetic, or flow proportional sampling are not rigidly adhered to. Normally, the sample is collected over one full cycle of each desired set of process operating conditions. When a series of discrete samples result, they are combined to produce a single "average" value for the total process cycle.

Level 2 sampling programs are directed toward a more detailed representation of stream composition. They are not as "inclusive" as Level 1, in that resources are expended to improve information only on streams of a critical nature and on compound classes defined as present by Level 1 analysis. Level 2 sampling is optimized for specific compounds or classes of compounds contained in the streams sampled. Level 2 also provides a more quantitative description of the concentrations and mass flow rates of the various substances in the stream. Further, recommended procedures for compliance testing should be introduced into the program to ensure that the data acquired by the procedures selected for Level 2 sampling can be correlated with regulatory requirements.

One primary refinement will be the need for more rigorous attention to selection and preparation of sampling sites. Additionally, adherence to procedures for acquiring a representative sample must be incorporated into the test procedures. Level 2 sampling should also provide for replication of samples in order to further improve accuracy and be more representative. In many cases, Level 2 sampling will require not only more care in use but also modifications of Level 1 equipment and/or the application of entirely new sampling methods. Such cases result from the necessity to more definitively identify the materials which produce the adverse environmental problems. For example, if Level 1 has indicated a high concentration of sulfur-containing species, Level 2 sampling must be specifically designed to isolate the sulfur-containing materials into appropriate fractions which can be analyzed for individual compounds or compound classes.

At Level 3, emphasis is placed on the variability of stream composition with time and process or control system parameters. When it has been determined at Level 1 that a stream is environmentally unacceptable and at Level 2 what the unacceptable components are, it is necessary to accurately define the range of values to be expected and the effectiveness of a control process if control equipment is installed. An effective Level 3 sampling program is designed to monitor a limited number of selected compounds or compound classes. Since Level 3 sampling is designed to provide information over a long period of time the sampling procedures must be carefully planned to minimize costs. Such programs must be tailored to the specific requirements of each stream being monitored. Based on the information developed at Level 2, specialized sampling procedures can be designed to track key indicator materials at frequent intervals. If at all possible, Level 3 should incorporate continuous monitors where appropriate.

During Level 3 programs, it is anticipated that Level 2 sampling will be conducted at predetermined intervals to check the limited Level 3 information. Further, recommended procedures for compliance testing should continue to be applied in the program to ensure that data correlations with regulatory requirements can be met on a long-term basis.

Analytical Methodology in a Phased Approach

During an environmental source assessment, the analytical methods applied will vary from relatively simple, manual wet chemistry to highly complex instrumental techniques. Analyses proceed from general, broadly applicable survey methods to more specialized techniques tailored to specific component measurements. This very broad range requirement has been structured to adhere to the same level concept described for the sampling program. At each phase of the analytical program, the depth and sophistication of the techniques are designed to be commensurate with the quality of the samples taken and the information required. Hence, expenditure of analytical resources on screening samples from streams of unknown pollution potential is minimized.

Level 1 sampling provides a single set of samples acquired to represent the average composition of each stream. This sample set is separated, either in the field or in the laboratory, into solid, liquid, and gas-phase components. Each fraction is evaluated with survey techniques which define its basic physical, chemical, and biological characteristics. The survey methods selected are compatible with a very broad spectrum of materials and have sufficient sensitivity to ensure a high probability of detecting environmental problems. Analytical techniques and instrumentation have been kept as simple as possible in order to provide an effective level of information at minimum cost. Each individual piece of data developed adds a relevant point to the overall evaluation. Conversely, since the information from a given analysis is limited, all the tests must be performed to provide a valid assessment of the sample.

Physical analysis of solid samples is incorporated into Level 1 because the size and shape of the particles have a major effect on their behavior in process streams, control equipment, atmospheric dispersion, and the respiratory system. In addition, some materials have characteristic physical forms which can aid in their identification.

Chemical analyses which determine the types of substances present, are incorporated to provide information for predicting control approaches, atmospheric dispersion/-transformation, and potential toxicity of the stream. Finally, because prediction of hazard based on physical and chemical analyses alone is subject to many uncertainties, biological assay techniques are incorporated as a measure of the potential toxicity. The basic Level 1 analytical procedures are given in Table II.

Table II

Basic Level 1 Analyses

Physical:	Cyclone particle size
	Optical microscopy
Chemical:	Spark source mass spectormetry
	Wet chemical (selected anions)
	Gas chromatography
	Liquid chromatography
	Infrared spectrometry
	Low resolution mass spectrometry
Biological:	Rodent acute toxicity
	Microbial mutagenesis
	Cytotoxicity
	Fish acute toxicity
	Algal bioassay
	Soil microcosm
	Plant stress ethylene

The analytical procedures applied at Level 2 may be extensions of the Level 1 procedures. In most cases, however, information developed at Level 1 will provide background for selection and utilization of more sophisticated sampling and analysis techniques. Because Level 2 analyses must positively identify the materials in sources which have already been found to cause adverse environmental effects, these analyses are the most critical of all three levels. It is equally important, however, that the analyses be conducted in an information-effective manner. This is because increasing specificity and accuracy result in cost escalations which are, at best, exponential rather than proportional. Due to the multiplicity of analytical techniques required and the potential for unnecessarily high expenditures, the analyses must be conducted with a full awareness of the information requirements of the environmental assessment program.

It is evident from the preceding comments that Level 2 analyses cannot be conducted via a prescribed series of tests. Each sample will require the analyst to select appropriate techniques based on the information developed in Level 1 and the information required for the assessments. Needless to say, this will encompass at one time or another virtually all the tools available to the analytical chemist and biologist, and a great deal of skill in their application. Of major importance will be the use of separation schemes for isolating fractions since the interpretation of both chemical and biological data from complex samples is a difficult, if not impossible, task. As with Level 2 sampling, the analytical scheme should incorporate the chemical and biological tests required by regulatory agencies.

The analytical procedures for Level 3 are specific to selected components identified by Level 2 analysis and are oriented toward determining the time variation in the concentrations of key indicator materials. In general, the analysis will be optimized to a specific set of stream conditions and will therefore not be as complex or expensive as the Level 2 methods. Both manual and instrumental techniques may be used, provided they can be implemented at the process site. Continuous monitors for selected pollutants should be incorporated in the analysis program as an aid in interpreting the data acquired through manual techniques. The total Level 3 analysis program should also include the use of Level 2 analysis at selected intervals as a check on the validity of the key indicator materials which reflect process variability.

Discussion

The major emphasis in environmental measurements over the past 2 years has been on development of procedures for Level 1. As presently conceived, Level 1 is structured to produce a cost-effective information base for prioritizing streams and for planning any subsequent programs. The specific procedures applied are detailed in two published manuals, the first covering sampling and chemical analysis (4) and the second biological tests (5). Application of the complete battery of procedures provides input data to support evaluation of the following questions:

a. Do streams leaving the processing unit have a finite probability of exceeding existing or future air, water, or solid waste standards or criteria?

b. Do any of the streams leaving the processing unit contain any classes of substances that are known or suspected to have adverse environmental effects?

c. Into what general categories (classes) do these substances fall?

d. What are the most probable sources of these substances?

e. Based on their adverse effects and mass output rates, what is the priority ranking of streams?

f. For streams exhibiting potential adverse environmental effects, what is the basic direction that control strategies are likely to follow?

The overall accuracy goal of Level 1 is to report the true value within a factor of 3. This means that a reported value between 30 and 300 is acceptable when the true value is 100. Many of the Level 1 sampling and analysis techniques are more accurate than this goal, but the relaxation of traditional limits allows significant cost savings. In general, grab sampling procedures are utilized for solid and liquid samples in order to minimize sampling costs.

In the case of particulate laden stack gases it is necessary to use sophisticated and relatively expensive equipment to collect valid samples, even at Level 1. To provide this type of unbiased sampling, the acquisition of gas samples requires the use of a sampling train designed specifically for environmental assessment sampling. The Source Assessment Sampling System (SASS) (6) has been designed to operate at 150 1pm and collect both solids and vapors. The entrained particulates are fractionated into four sizes: $> 10\mu m$, 3-10 μm, 1-3 μm, and < 1 μm. Vapor-phase organic materials are adsorbed on a solid sorbent (XAD-2), and the inorganic vapors are retained in the chemically active impinger solutions. Because it is impossible to predict the concentration of a constituent in the gas stream at the start, sampling is based on the minimum volume of gas necessary to provide detection of materials in the analytical scheme. A minimum of 30 m^3 (1000 ft^3) is sufficient to ensure detection of materials in the source at 1 mg/m^3 (approximately 1 μg/m^3 in the receiving media). The overall process scheme for samples from this system is shown in Figure 2.

Sampling of fugitive air emissions presents problems similar to those of stack gases and specialized equipment is also necessary for this operation. A Fugitive Assessment Sampling Train (FAST) has been designed which operates at 5.2 m^3/min. Entrained particles are fractionated into > 15μm, 3-15 μm, and < 3 μm and organic vapors from a side stream are trapped in an XAD-2 sorbent trap.

An overview of the Level 1 chemical analysis scheme is shown in Figure 3. If divided along classical lines of inorganic and organic analysis, it can be seen that the primary techniques applied for inorganic analysis are gas chromatography for the gaseous components and spark source mass spectrometry (SSMS) for elemental analysis of solids and liquids. A number of anions including sulfate, nitrate, and phosphate are presently analyzed by wet chemical techniques. However, future analysis will be based on ion chromatography to increase the number of anions detectable. Atomic absorption has been applied only for mercury, arsenic, and antimony which were thought to be lost by vaporization during SSMS analysis. Recent data indicates that SSMS values for arsenic are within Level 1 accuracy requirements and that antimony values may also be acceptable. It is probable that AA will be retained in future schemes only for mercury analysis. Further development is also necessary to improve the comprehensiveness of inorganic gas analysis.

The Level 1 organic analysis strategy employs four analytical operations in the control scheme. Gas chromatography is applied at the sampling site to analyze for organic substances with boiling points below 100°C. The separation is essentially by boiling point with flame ionization detection. Column conditions are such that individual compound separation is not achieved — the desired goal being to group all materials into seven boiling point ranges.

In the laboratory conducted portion of the organic analysis, a seven step liquid elution chromatography separation on silica gel forms the central part of the scheme. It is an analytical step (in that behavior of a given class of compounds is predictable) as well as a separation step (since the fractions may be further analyzed much more readily than the original mixture). The behavior of selected classes of compounds with respect to the chromatographic conditions employed is shown in Figure 4.

The second laboratory operation is determined of the total organic content. This operation allows quantitation of the organics in each of the chromatographic fractions as well as aliquot size selection for optimum column operation. The original Level 1 scheme depended entirely upon reduction to dryness and weighing for total organics determination. Recent data show that many materials in the boiling range below 275°C may be partially lost by that approach. Accordingly, a gas chromatographic procedure for volatile organics has been adopted as a part of the Level 1 strategy. Total organic content is obtained by addition of the gravimetric results and the total chromatographable organics (TCO).

The third analysis is by infrared absorption spectrophotometry. This classical technique is often overlooked in today's mass-spectrometry dominated laboratory, but still remains a powerful tool which provides considerable information at moderate cost. Infrared spectra of the seven chromatographic fractions may be used to confirm

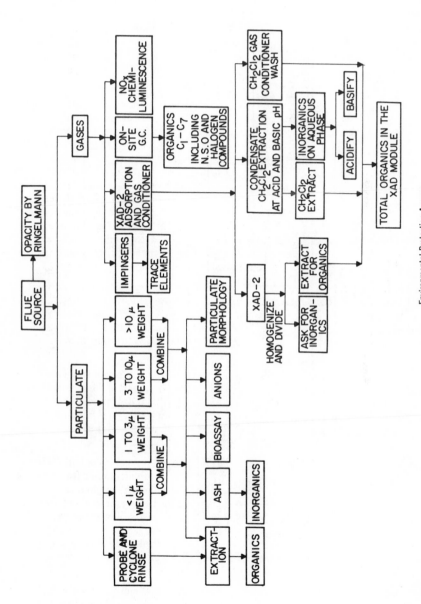

Environmental Protection Agency

Figure 2.   Flue gas sampling flow diagram (2)

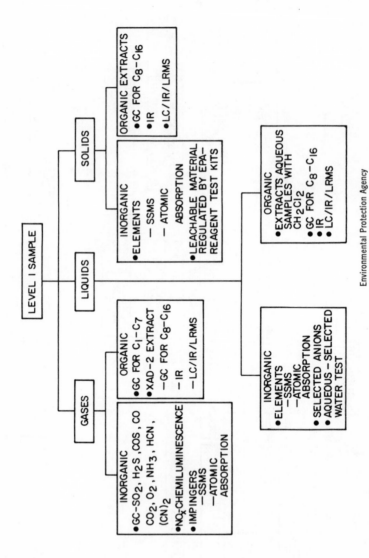

Figure 3. *Multimedia analysis overview* (2)

Environmental Protection Agency

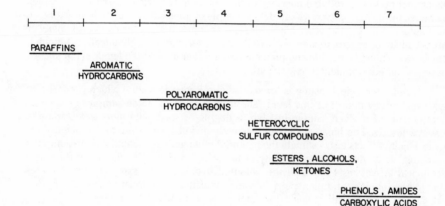

Figure 4. Liquid chromatographic fractions vs. class types

the presence or absence of compound classes or functional groups as indicated by the chromatographic data. It is occasionally possible to identify specific compounds from the infrared spectra but, as previously mentioned, the complexity of most environmental samples makes this the exception rather than the rule.

Finally, low resolution mass spectrometry (LRMS) is applied to each separation fraction which exceeds a specified concentration when referred to the sample sources. The present levels are 0.5 $\mu g/m^3$ for gas streams, 1 mg/kg for solids, and 0.1 mg/1 for liquids. The selection of this particular technique, firmly in the middle of a transition between Levels 1 and 2, caused many philosophical problems concerning its proper use. The original Level 1 scheme did not contain LRMS. It has been included in the modified strategy to prevent potential triggering of Level 2 efforts based on large amounts of suspicious, but innocuous, organics. However, LRMS can be costly if constraints are not placed on how frequently it is applied.

Biological tests are incorporated in Level 1 because physical and chemical analysis cannot provide a reliable measure of the potential biological response to complex samples. In addition, bioassays can detect complex biological effects such as synergism and antagonism. On the other hand, bioassays cannot identify the cause of toxicity or mutagenicity or suggest means of controlling it. Thus, physical, chemical, and biological analyses must be used to complement one another at all three levels of the phased approach to environmental assessment.

Level 1 biological testing is limited to whole sample testing which is consistent with the survey nature of this level. The testing of fractionated samples or specific components of a given sample involves a degree of specificity more appropriate to Level 2 testing. The bioassays include assessments of both health and ecological effects. Health effects tests estimate the potential mutagenicity, potential or presumptive carcinogenicity, and potential toxicity of the samples to mammalian organisms. The ecological effects tests focus on the potential toxicity of the samples to vertebrates (fish), invertebrates, and plants in freshwater, marine, and terrestrial ecosystems. Figure 5 shows the overall biological scheme applied to Level 1 samples.

Three primary tests are incorporated in the health effects area. The microbial mutagenesis assay is based on the property of selected *Salmonella typhimurium* mutants to revert from a histidine requiring state to prototrophy due to exposre to various classes of mutagens. The test can detect nanogram quantities of mutagens and has been adapted to mimic some mammalian metabolic processes by the addition of a mammalian liver microsomal fraction. The test is used as a primary screen to determine the mutagenic activity of complex mixtures or component fractions.

Cytotoxicity assays employ mammalian cells in culture to measure cellular metabolic impairment and death resulting from exposure *in vitro* to soluble and particulate toxicants. Mammalian cells derived from various tissues and organs can be maintained as short term primary cultures or, in some cases, as continuous cell strains or lines. The cytotoxicity assays, incorporated as part of Level 1 analysis, employ primary cultures of rabbit alveolar (lung) macrophages (RAM) and maintenance cultures of strain WI-38 human lung fibroblasts.

The third health effect bioassay employed utilizes acute toxicity in whole animals (rats). Since the major objective of the Level 1 biological testing procedure is to identify toxicology problems at minimal cost, a two-step approach is taken to the initial acute *in vivo* toxicology evaluation of unknown compounds. The first is based on a

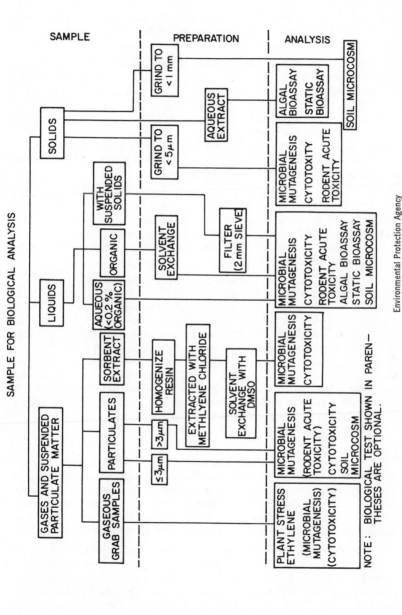

Environmental Protection Agency

Figure 5. Biological analysis overview (5)

quantal (all-or-none) response and the second on a quantitative (graded) response. Normally, the quantal test is used to determine the necessity to carry out the quantitative assay. In the quantal test, a single dose of 10 g/kg body weight of the sample being evaluated is administered to 10 rats and their behavior observed for 14 days. If no mortality occurs, additional tests are not conducted. If the administration of the sample causes a single death, a quantitative study on 80 animals is conducted at various dilutions of the sample.

There are also three primary tests applied to aquatic assays of water effluents. The same basic tests are used for both fresh water and marine assays; the primary variation is in the species selected. An algal assay is based on the principle that growth is limited by the nutrient that is present in shortest supply with respect to the needs of the organism. The test is designed to be used to quantify the biological response (algal growth alteration) to changes in concentrations of nutrients and to determine whether or not various effluents are stimulative or inhibitory to algal growth. Measurements are made by adding a selected test alga to test water containing the sample and determining algal growth rates at appropriate intervals.

Acute static bioassays are employed to evaluate sample effects on fish and micro-invertebrates. Fathead minnows and *Daphnia pulex* are the freshwater species employed, while sheepshead minnows and grass shrimp are used for marine assays. In all cases, the test organisms are exposed for 96 hours to prescribed concentrations of the sample introduced into holding tanks maintained at the same environmental conditions as a control population.

Two bioassays are employed to evaluate the effect of samples on terrestrial life forms. For gas samples, the plant stress ethylene test is presently recommended. This test is based on the well-known plant response to environmental stress: release of elevated levels of ethylene (under normal conditions plants produce low levels of ethylene). The test is designed to expose plants to various levels of gaseous effluents under controlled conditions. The ethylene released during a set time period is then measured by gas chromatography to determine toxicity of the effluent. For liquid and solid samples, a soil microcosm test is employed. The sample is introduced on the surface of a 5 cm diameter by 5 cm deep plug of soil obtained from a representative ecosystem. Evolution of carbon dioxide, transport of calcium, and dissolved oxygen content of the leachate are the primary quantifying parameters.

Results

All the elements in the Level 1 scheme have been utilized in studies of discharges from industrial and energy processes. The data developed is far too voluminous to be presented here and the results discussed are, therefore, more a demonstration of the utility and viability of selected elements within the scheme. It should also be noted that all the data presented was acquired to evaluate the Level 1 methodology and should *not* be interpreted as indicative of emissions from the processes studied.

In a study of the precision of the SASS train for the collection of particulates and organics from gas streams, total particulate loadings determined with two independent SASS trains agreed within 20 percent of the value determined with a standard EPA Method 5 train operated simultaneously on triplicate runs. The size distributions determined by the two trains were also in good agreement for all three runs as illustrated in Figure 6. The quantity and distribution of extractable organics between the two systems was also well within Level 1 tolerances as shown in Table III.

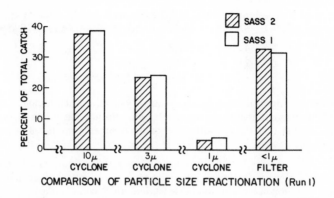

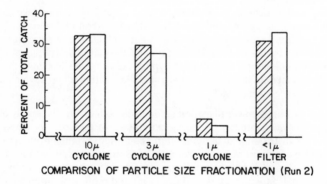

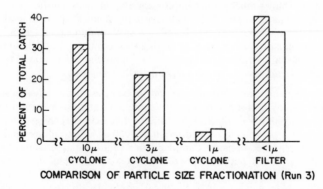

Figure 6.   *Comparison of particle size fractionation of two SASS trains*

Table III
Organic Extractables (mg/m^3)

	Cyclone		XAD-2 (Extract)		XAD-2 (Module)	
	SASS-1	SASS-2	SASS-1	SASS-2	SASS-1	SASS-2
TCO	0.03	0.01	3.4	3.6	(Rinse)	
Grav	1.6	1.6	10.2	9.0	69	81
Total	1.7	1.6	13.6	12.6	69	81

Additional analysis of these samples using the liquid chromatographic separation, total chromatographable organics, infrared, and low resolution mass spectrometry demonstrated that the sampling and analysis scheme produced reproducible data as shown in Table IV.

The samples acquired from these tests were also used to evaluate inter- and intra-laboratory precision for the various elements of the anaytical scheme. In general, it was found that the analysis for total extractable orgaic content and the distribution between chromatographable and gravimetric fractions were much better than the factor of 3 required by Level 1. Triplicate analysis by four laboratories gave an average total organic content of 480 mg and a coefficient of variation of 4 percent. However, the results of the liquid chromatographic separation indicated a wider range of values in the distributions by fractions, indicating a need for more carefully controlled column preparation and sample handling. A typical set of values is shown in Table V. (This and subsequent data was acquired with an eight fraction scheme which has since been reduced to seven fractions.)

Tests of a coal fired power plant and an oil fired boiler show the ability of Level 1 to discriminate between process operating conditions. In Figure 7, the cross hatched bars represent the distribution of organics under conditions producing maximum operating temperatures. Under these conditions, the maximum weight of organic emissions was found in fraction 5, with generally much smaller concentrations in all other fractions. With water injection on the oil fired unit and reduced load on the coal fired unit, the organics show a much more complex distribution with significant increases of materials in fractions 2, 3, and 7. In both cases, the concentration of total organics detected in the emissions by the Level 1 analyses was found to increase by a factor of 2 under the conditions producing lower operating temperatures. While these results do not identify the specific compounds present, they show the magnitude of the change and the classes of compounds which warrant further analysis.

An indication of the utility of bioassay procedures in the Level 1 scheme is available from a study performed on the particulate collected from 10 industrial process streams prior to control devices (7). An initial cytotoxicity screening of the samples was performed at a particulate concentration of 1000 μg/ml of culture medium containing rabbit aveolar macrophages. Samples found in the initial screening to produce cell deaths greater than 15 percent were retested at concentrations of 1000 μg/ml, 300 μg/ml, and 100 μg/ml of culture medium. In addition, an attempt was made to ascertain whether the toxicity of a given particulate sample was due to the particles themselves and/or soluble component(s) released into the medium. The particulate matter was incubated in the culture medium (less serum but including antibiotics) for 20 hours and then removed from the medium via centrifugation and filtration through a 0.22 μm Millipore filter. The filtered supernatant and centrifuged particles (resuspended in

TABLE IV. Organic Categories in SASS Samples

Categories	Concentration (mg/m³)	
	SASS-1	SASS-2
Aliphatic hydrocarbons	1.1	1.1
Halogenated aromatic HC's	—	—
Aromatic HC's-benzene	0.6	0.1
<216	28.7	28.2
>216	25.6	28.7
Heterocyclic N	20.1	24.0
Heterocyclic S	2.4	2.5
Heterocyclic O	2.2	6.7
Phenols	0.2	0.3
Esters	0.5	0.2
Ethers	—	—
Amines	—	—
Amides	—	—
Carboxylic acids	0.6	0.7
Sulfonic acids, sulfoxides	—	—
Sulfur	0.2	0.7
Inorganics	0.1	—
Unclassified	0.3	1
Silicones	0.1	—

TABLE V. TCO/GRAV Results: Field Sample, XAD-2 Extract, Run 1 (mg)

Fraction	TCO				GRAV				Total			
	A	B	C	D	A	B	C	D	A	B	C	D
LC1	1.4	<0.4	12.0	2.1	11.5	4.0	4.8	––	13	4	17	2
LC2	15.2	<0.4	60.3	120	4.6	6.4	84.2	200	20	7	145	320
LC3	54.7	31	40.3	2.6	259.4	284	177.6	19	314	315	218	22
LC4	0.5	2.3	9.2	1.4	10.1	22	12.8	––	11	24	22	1
LC5	0.9	0.84	0.4	7.4	31.3	8.8	7.6	––	32	10	8	7
LC6	3.2	1.0	18.4	13	16.6	38	27.1	110	20	39	46	123
LC7	0.5	<0.4	21.6	1.9	2.8	23	3.6	––	3	23	25	2
LC8	0.9	<0.4	N.R.	8.3	3.7	37	0	50	5	37	0	58
TOTAL	77.3	36.7	162.2	156.7	340	423.2	317.7	379	418	459	481	535

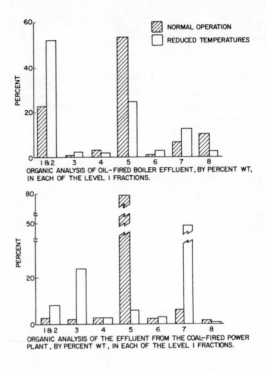

Figure 7.   Effect of process variables on organic weights in Level 1 fractions

TABLE VI.   Viability of Rabbit Alveolar Macrophages
Exposed to Sub-Micron Particle Filters

Filter Sample	Total Sample	Supernatant	Dried Filter
Oil Fired Power Plant ($<3\mu$m)	20.4[a]	10.5[a]	30.1[a]
Copper Smelter ($<3\mu$m)	26.3	29.5	41.0
Aluminum Smelter ($<3\mu$m)	28.2	24.9	86.3
Aluminum Smelter ($>3\mu$m)	31.5	32.1	60.8
Iron Sintering ($<3\mu$m)	33.0	55.2	95.7
Copper Smelter ($>3\mu$m)	37.2	30.6	35.6
Iron Sintering ($>3\mu$m)	56.1	58.8	99.5
Paper Mill ($<3\mu$m)	56.1	76.7	97.9
Sludge Incinerator ($<3\mu$m)	92.1	71.8	93.8
Ceramics Plant ($<3\mu$m)	93.0	53.6	98.4
Open Hearth Furnace (Total)	95.9	98.8	97.4
Coke Oven Heaters ($<3\mu$m)	97.6	98.8	99.6
Basic Oxygen Furnace (Total)	98.3	98.4	99.6
Teflon Control Filter	Non-Toxic	Non-Toxic	Non-Toxic

[a]Percent viable cells (mean of two observations)

fresh medium) were then independently tested for cytotoxicity. Table VI is a tabulation of the viability (percent living cells) of the samples tested. The sources have been ranked on a relative basis of toxicity of the total sample. Except for the aluminum smelter samples, the finer particulate (< 3 μm) is more toxic than the larger (> 3 μm). Comparing the second and third columns (Supernatant and Dried Filter) indicate that the leachable components (supernatant) were usually responsible for the observed toxicity. In the case of the oil fired power plant and copper smelter samples, however, the undissolved particulate retained a significant level of toxicity.

Conclusions

The phased approach developed for environmental assessments has been successfully implemented on a number of programs. The Level 1 results have been useful for defining potential problems for further study and directing additional resources into the areas of greatest need. While not all the elements of the scheme have been found effective on each study, every element has produced needed data in at least one or more of the individual studies. In considering the utility of the approach it is therefore necessary to consider the overall results rather than the value of any single component.

The procedures selected initially have been constantly improved in detail; however, no significant changes in the basic techniques have been necessary. The phased approach is, therefore, a viable and cost effective concept for identifying potential environmental problems associated with discharge from industrial and energy processes.

Acknowledgements

The people making significant contributions to the development of the phased approach are too numerous to list here. We trust that they, as well as the reader, will recognize that this paper would not have been possible without their efforts.

Literature Cited

1. Dorsey, J. A., Johnson, L. D., Statnick, R. M. and Lochmuller, C. H., "Environmental Assessment Sampling and Analysis: Phased Approach and Techniques for Level 1." U. S. Environmental Protection Agency Report, EPA-600/2-77-115, NTIS No. PB 268-563/AS (6/77).

2. Hammersma, J. W. and Reynolds, S. L., "Field Test Sampling/Analytical Strategies and Implementation Cost Estimates: Coal Gasification and Flue Gas Desulfurization." U. S. Environmental Protection Agency Report, EPA-600/2-76-093b, NTIS No. PB 254-166/AS (4/76).

3. Vlahakis, J. and Abelson, H., "Environmental Assessment Sampling and Analytical Strategy Program." U. S. Environmental Protection Agency Report, EPA-600/2-76-093a, NTIS No. PB 261-259/AS (5/76).

4. Hammersma, J. W., Reynolds, S. L. and Maddalone, R. F., "IERL-RTP Procedures Manual: Level 1 Environmental Assessment." U. S. Environmental Protection Agency Report, EPA-600/2-76-160a, NTIS No. PB 257-850/AS (6/76).

5. Duke, K. M., Davis, M. E. and Dennis, A. J., "IERL-RTP Procedures Manual: Level 1 Environmental Assessment, Biological Tests for Pilot Studies." U. S. Environmental Protection Agency Report, EPA-600/7-77-043, NTIS No. PB 268-484/AS (4/77).

6. Blake, D. E., "Source Assessment Sampling System: Design and Development." U. S. Environmental Protection Agency Report, EPA-600/7-78-018, NTIS No. PB 279-757/AS (2/78).

7. Mahar, H., "Evaluation of Selected Methods for Chemical and Biological Testing of Industrial Particulate Emissions." U. S. Environmental Protection Agency Report, EPA-600/2-76-137, NTIS No. PB 257-912/AS (5/76).

RECEIVED November 17, 1978.

The Identification and Measurement of Volatile Organic Compounds in Aqueous Environmental Samples

THOMAS A. BELLAR, WILLIAM L. BUDDE, and JAMES W. EICHELBERGER

Environmental Protection Agency, Office of Research and Development, Environmental Monitoring and Support Laboratory, Cincinnati, OH 45268

During the early 1970's the increasing concern over the potentially harmful effects of synthetic organic compounds in the environment led the Environmental Protection Agency (EPA) to begin to develop laboratory analytical methods for organic pollutants. In the area of natural and wastewater analyses, the logical first approach to the isolation and concentration of organic pollutants was the time honored liquid-liquid extraction methods. Liquid-liquid extraction had been employed for many years in analytical chemistry, and was fairly well developed for the analysis of some chlorinated hydrocarbon pesticide residues. During this period it was also recognized that compound separation and detection methods would be required that were not only responsive to a wide variety of organic compositions and structures, but also gave sufficient qualitative information to permit unequivocal identification of the compounds observed. Thus the seeds were planted for the intensive development of computerized gas chromatography-mass spectrometry (GC/MS). The solvent extraction and GC/MS procedures were naturally compatible, and the overall result was a significant advancement in organic analytical methodology.

During this development period there was the clear recognition that the liquid-liquid extraction — GC/MS methodology had several significant deficiencies. On one hand, there was the vast number of polar, water soluble organic compounds that were not partitioned efficiently into the organic solvent. Another concern was the large number of compounds that were extracted, but were not amenable to gas chromatography because of insufficient vapor pressures at even elevated temperatures. At the other extreme was the large number of very volatile compounds that were efficiently partitioned into a wide vareity of organic solvents, but were not observed because they were either lost during extract concentration, or masked during solvent elution from the GC. In early 1973, insufficient resources were available to attack all of these problems simultaneously, but a program was started at the EPA'S Environmental Research Center in Cincinnati to develop methods for the identification and measurement of immiscible organic solvents and similarly volatile compounds in water samples.

Four general approaches were considered for this methodology:
1. Direct aqueous injection gas chromatography.
2. Use of an extremely volatile solvent for liquid-liquid extraction that would permit chromatographic resolution of other volatile constituents.

3. Use of a high molecular weight, late eluting solvent that would not mask early eluting volatile compounds.
4. Use of some type of gas purging and adsorbent trapping system that would eliminate solvent interferences and permit a high degree of concentration.

Direct aqueous injection GC/MS was investigated [1] using both continuous repetitive measurement of spectra (CRMS) and selected ion monitoring (SIM). These GC/MS data acquisition methods, which have been defined in some detail [2], differ in sensitivity by three to four orders of magnitude. With CRMS, which generates complete mass spectra for compound identification, a lower detection limit of about 1 milligram/liter was established. While this is acceptable for many waste samples, it is much too high for analyses of surface or drinking water. The application of SIM provides adequate detection limits, but imposes the constraint that one must know what compounds are to be measured before the analysis in order to select the ions that must be monitored in real time. Also with SIM, the full mass spectra are not available for the identification of the ubiquitous unexpected compounds.

The use of a volatile solvent, e.g., pentane, was not explored because of inherent limitations. Concentration of such extracts was not possible because of the volatility of the sample components. Therefore the maximum concentration factor that could have been achieved was limited by the partition coefficients of the compounds into the solvent used in the extraction. For most compounds this factor was estimated to be about 10:1. Furthermore, with CRMS and other general detectors, the solvent masking problem would still preclude observation of many compounds. Therefore, the method would be limited to detectors that are not responsive to the solvent used in the extraction. Recent work [3,4,5] has indicated that extraction with a volatile solvent is a viable approach for the analysis of a small set of compounds, e.g., the trihalomethanes, with an electron capture detector in drinking water samples where concentration factors of 10:1 or less are acceptable.

The application of a heavy, late eluting solvent was explored somewhat [6], using decalin and hexadecane solvents. While this method did show some promise, it was not investigated extensively because of limited resources.

The concept of using a gas purging and adsorbent trapping system was considered very promising, and was explored extensively for several reasons. The method appeared to be quite general and would be amenable to use with almost any gas chromatographic detector, including mass spectrometry, flame ionization, electron capture, electrolytic conductivity, etc. In addition, the method would afford a very important preliminary separation of very volatile compounds from often very messy samples prior to gas chromatography. This feature would allow general application of the method to samples as diverse as drinking water and raw undiluted sewage. Finally, earlier unpublished [7] work with air samples had shown that suitable adsorbents were available and thermal desorption was feasible.

The balance of this paper is concerned with a presentation of the details of the gas purging and adsorbent trapping method for the analysis of very volatile compounds in water samples. A number of method variables have been studied during the last five years, and the method has been applied to a wide variety of sample types. There have been a number of publications which are cited and may be consulted for additional information [8-12].

Method Definition and Apparatus

Gas purging and trapping is a method for the isolation, concentration, and determination of low boiling organics in water. The method uses finely divided gas bubbles passing through the water sample to transfer organic compounds from the aqueous to the gas phase. The compounds concentrate by adsorption on a porous polymer trap at room temperature as the purge gas is vented. The compounds are subsequently desorbed at elevated temperature by backflushing wth a carrier gas into the gas chromatographic system. The method may provide both qualitative and quantitative information. Purging may be accomplished at ambient or elevated temperatures with helium or another inert gas. A mass spectrometer or other type of detector may be used.

The equipment required for this method consists of a purging device, a trap, and a trap heater or desorber. Figure 1 shows construction details for an all glass purging device with a 5 ml sample capacity. The glass frit at the base of the sample volume allows the finely divided gas bubbles to pass through the sample while the sample is restrained above the frit. Gaseous volumes above the sample are kept to a minimum to eliminate dead volume effects, yet sufficient space is allowed to permit most foams to disperse. The inlet and exit ports are constructed of heavy walled quarter inch glass tubing to permit leak free removable connections with finger tight compression fittings containing Teflon furrules. The removeable foam trap is optional and recommended for samples that foam. A 25 ml capacity purging device is recommended for use with a mass spectrometer GC detector used in the CRMS mode.

Figure 2 shows a trap which is a short gas chromatographic column that retards the flow of the compounds of interest at ambient temperature while venting the purge gas and, depending on the adsorbent used, much of the water vapor. The trap is constructed with a low thermal mass to allow rapid heating for efficient desorption, and rapid cooling to ambient temperature for recycling. The trap length, diameter, and wall thickness indicated in Figure 2 are critical and variations in these will affect the trapping and desorption efficiencies of the compounds discussed in this paper.

The trapping and desorption efficiencies are also a function of the adsorbents, adsorbent mass, and the adsorbent packing order shown in Figure 2. The single adsorbent Tenax GC (60/80 mesh) is effective for compounds that boil above approximately 30°C. However compounds that boil below approximately 30°C are not strongly adsorbed by Tenax and may be vented under the purging conditions. If compounds that boil below about 30°C are to be measured, a dual adsorbent trap should be used. Grade-15 silica gel effectively retards the flow of most organics at ambient temperature and should be packed behind the Tenax to trap the lower boiling components. Silica gel is not a useful single adsorbent because higher boiling compounds do not efficiently desorb from it at 180°C.

The Tenax-silica gel combination trap utilizes the adsorptive properties of two materials to provide a trap that effectively adsorbs and desorbs a wide variety of organic compounds. The small amount of OV-1 on glass wool at the trap inlet (Figure 2) is to insure that all the Tenax adsorbent is within the heated zone and is efficiently heated to the desorption temperature. A metal fitting at the trap inlet could act as a heat sink and create a cool spot on the Tenax if this spacer is not used.

Details of the trap heater are also shown in Figure 2. The adsorption-desorption cycle may be accomplished conveniently with the use of a six port valve and a plumbing system constructed of materials that neither adsorb volatile organics nor outgas

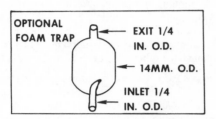

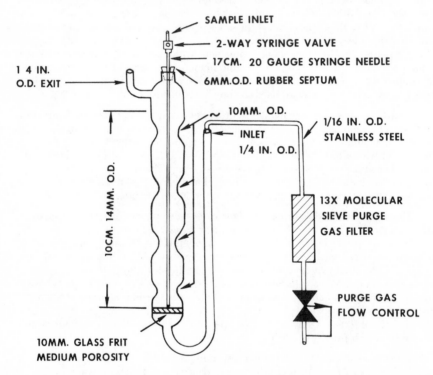

Figure 1.   A purging device with a 5-mL sample capacity

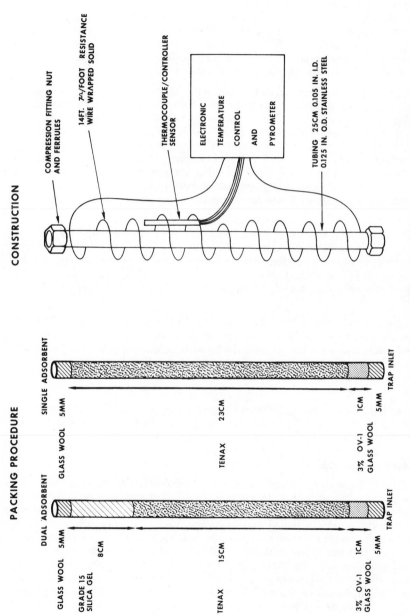

Figure 2. Trap assembly for a purging device

them. Several commercially available purge and trap systems use this approach. With the six port valve in the adsorb position, the effluent from the purging device passes through the trap where the flow of the organics is retarded and the purge gas is vented. During this period the gas chromatograph is supplied with carrier gas and may be used for other analyses. With the valve in the desorb position, the trap is placed in series with the gas chromatographic column which allows the carrier gas to back flush the trapped materials onto the chromatographic column.

It is strongly recommended that the power for the desorber heater be supplied by an electronic temperature controller that is set to begin supplying power as the valve is placed in the desorb position. This allows rapid heating of the trap to 180°C with minimal overshoot and maintenance of the desorb temperature until desorption is complete (a four minute backflush at 20-60 ml/min is recommended). Using this procedure, the trapped compounds are released as a narrow plug into the gas chromatograph, which should be at the initial operating temperature. Packed columns with theoretical efficiencies near 500 plates/foot under programmed temperature conditions can usually accept such desorb injections without altering peak geometry.

Substitution of a non-controlled power supply, such as a manually operated variable transformer, will cause non-reproducible retention times and may lead to unreliable concentration measurements. If it is not possible to heat the trap in a rapid and controlled manner to the desorption temperature, the contents of the trap may be transferred onto the analytical column at 30°C or lower and once again trapped. The analytical column is then rapidly heated to the initial operating temperature for the analysis.

Several gas chromatographic columns have been employed for the separation of the volatile components prior to measurement. A recommended column for general purpose work is a 8-ft x 0.1 -in. id stainless steel or glass tube packed with 0.2% Carbowax 1500 on Carbopack-C (80/100 mesh). With a helium flow of 40 ml/min., the initial temperature of 60°C is held for three min., then programmed at 8°C/min. to 160°C.

Newly packed traps should be conditioned overnight at 230°C with an inert gas flow of at least 20 ml/min. The trap is also conditioned prior to daily use by backflushing at 180°C for 10 min.

Discussion of The Method

Table I contains a list of some of the compounds that have been submitted to this type of analysis. The recovery data is intended to be illustrative only since recoveries depend strongly on several important method variables. Recoveries are expressed as a percentage of the amount added to organic free water. The purge time was 11-15 minutes with helium or nitrogen, the purge rate was 20 ml/minute at ambient temperature, and the trap was Tenax followed by Silica Gel. Data from the 5 ml sample was obtained with a custom made purging device and either flame ionization, microcoulometric, or electrolytic conductivity GC detectors. Data from the 25 ml sample was obtained with a Tekmar commercial liquid sample concentrator and a mass spectrometer GC detector using CRMS.

A number of other compounds have been concentrated and measured using the purge and trap method, but no recovery data for these is available. These compounds include chloromethane, bromoethane, chloroethane, 1,2-dichloropropane, trans-1,3-dichloropropene-1, cis-1,3-dichloropropene-1,

TABLE I. Recoveries of Organics by Gas Purging and Trapping

Compound Added	Amount Added ug/l	% Recovery 5 ml Sample	Amount Added ug/l	% Recovery 25 ml Sample
Chlorinated Hydrocarbons				
methylene chloride			12	97
chloroform	160	101	12	95
chlorodibromomethane	8,2	86,69	12	72
bromodichloromethane	40,2	94,82	12	72
carbontetrachloride	4,2	108,82		
1,2-dichloroethane	2	70	12	95
1,1,2-trichloroethylene	2,2	118,89		
tethrachloroethylene	2,2	81,80		
chlorobenzene	4	80		
p-dichlorobenzene	4	71	12	58
dichlorofluoromethane	2	7		
trichlorofluoromethane	2	99		
vinyl chloride	2	95		
1,1-dichloroethylene	2	114		
1,1-dichloroethane	2	90		
trans-1,2-dichloroethylene	2	100		
1,1,1-trichloroethane	2	88		
dichloroiodomethane	2	72		
2,3-dichloropropene-1	2	85		
Hydrocarbons				
n-pentane	81	100		
n-nonane	93	99		
n-pentadecane	100	56		
benzene			12	105
toluene			12	91
Brominated Hydrocarbons				
bromoform	4,2	67,48	12	73
1,2-dibromoethane	4,2	80,47		
Nitrogen Compounds				
nitromethane			24	2
nitrotrichloromethane			24	22
N-nitrosodimethylamine			24	0
N-nitrosodiethylamine			12	0
N-nitrosodi-n-butylamine			12	0

1,1,2-trichloroethane, 2-chloroethylvinylether, 1,1,2,2-tetrachloroethane, and ethylbenzene. In general the method is applicable to compounds that have a low solubility in water and a vapor pressure greater than water at ambient temperature.

All the experiments summarized in Table I were conducted with the water sample at ambient temperature, about 22°C. Purging at elevated temperatures has been investigated (13) and clearly affects recoveries of some compounds. However, this is not recommended as a general procedure because the elevated temperature may promote chemical reactions that could significantly alter the composition of the trace organics. This is especially important with chlorinated drinking water or waste effluents, a subject that is discussed later in this paper.

A significant method variable is the purge gas flow rate. The total purge time is not very flexible since it will usually be desirable to keep this as short as possible to minimize the analysis time. In all of the experiments summarized in Table I, a flow rate of 20 ml/min. was employed for 11-15 minutes, and for many compounds an acceptable recovery was obtained. Figure 3 shows the percentage recovery of several representative compounds as a function of purge gas flow rate. The general curve shape displayed for 1,2-dichloroethane and bromoform is typical of most of the compounds in Table I. The low boiling compound vinyl chloride was trapped on Tenax only and its flow rate curve illustrates the sharp reduction in trapping efficiency observed with this type of compound and trap at elevated flow rates. The compound dichlorodifluoromethane displays a similar flow rate curve even with the combination Tenax-silica gel trap.

Because of the differences in the construction of various purge and trap devices, actual recoveries may vary significantly from those shown in Figure 3 and Table I. Therefore it is required that individual investigators determine recoveries of compounds to be measured as a function of flow rate with their apparatus. Operation in the optimum flow rate range will assure maximum sensitivity and precision for the compounds measured.

The recoveries of aliphatic hydrocarbons were found somewhat more variable than the recoveries of the other compounds investigated with this method. In all of these experiments, known quantities were added to organic free water, and the slightly soluble aliphatic hydrocarbons probably formed a thin surface layer on the water. Under these inhomogeneous conditions, special care is needed to achieve consistent results. A multiple phase equilibration head space method has been reported to give good results with aliphatic hydrocarbons [14].

In order to generate quantitative measurements within a reasonable purge time, e.g., 10-15 min., calibration of the method with known standards is required. The recommended approach is to estimate the concentration of the unknown by a comparison of its peak size with the size of the corresponding peak in a quality control check standard made at some appropriate concentration level and measured at regular intervals during the work day. From this estimate a concentration calibration standard is prepared with the concentrations of the compounds to be measured within a factor of two or less of the probable concentration in the unknown. Standards are prepared by taking aliquots of solutions in methanol and injecting them into organic free water. This insures maximum dispersal of the organic compound in the aqueous system. Organic free water is prepared by passing distilled water through an activated carbon column.

The standard is purged and measured immediately after the unkown and under the identical conditions used with the unknown. A sample of organic free water should be purged between each set of samples and especially after all high level standards or samples. This will insure that the apparatus is purged of contaminants and prevent cross contamination.

Sample matrices significantly different than surface water have not been investigated extensively. Extremes of pH, high ionic strength, or the presence of miscible

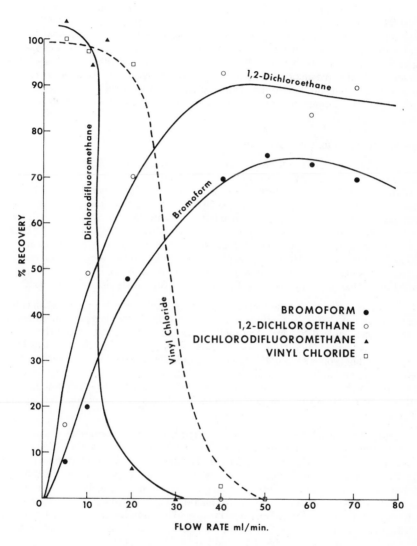

Figure 3. Recoveries of selected compounds as a function of purge gas flow rate

organic solvents will likely affect recoveries of some compounds. Therefore measurements of compounds in these matrices must include determinations of recoveries of spikes in the sample matrix and perhaps use of a calibration based on the method of standard additions.

The detection limit of the method is also dependent on a number of operational variables. For a sample volume of 5 ml a concentration factor of about 1000 over a direct aqueous injection is usually possible. This places the limit of detection in the 0.1 to 1 microgram per liter range for many detectors, including a GC/MS system operating in the SIM mode. With a GC/MS system operating in the CRMS mode, a sample volume of 25 ml is recommended to achieve this detection limit. The method can be applied over a concentration range of approximately 0.1 to 1500 micrograms per liter. Figure 4 shows a chromatogram of a mixture of 29 compounds from a purge and trap analysis.

Sample Collection and Preservation

Previous reports (12, 13) emphasized the importance of sample handling, and indeed because of the very volatile nature of the compounds measured in this type of analysis, sample collection deserves special consideration. In general, narrow mouth glass vials with a total volume in excess of 50 ml are acceptable. The bottles need not be rinsed or cleaned with organic solvents, but simply cleaned with detergent and water, rinsed with distilled water, air dried, and dried in a 105°C oven for one hour. The vials are carefully filled with sample to overflowing (zero head space) and a Teflon faced silicone rubber septum is placed Teflon face down on the water sample surface. The septa may be cleaned in the same manner as the vials, but should not be heated more than one hour because the silicone layer slowly degrades at 105°C.

Two types of seals for the vials have been employed and both give satisfactory results. Aluminum, one-piece, crimp-on seals used with serum vials and Teflon faced septa are acceptable if the seal is properly made and maintained during shipment. However, several years of experience indicates a success rate significantly less than 100% in making proper seals of this type in the field. Therefore simple screw cap vials used with the Teflon faced septa were evaluated and found to give equivalent results and a very high rate of acceptable samples. Narrow mouth screw cap bottles with Teflon-faced silicone rubber septa cap liners are strongly recommended for sample collection.

One special problem in sample preservation has been recognized (13) as a result of the widespread application of this method to drinking waters which contain residual quantities of disinfectants, e.g., chlorine. The levels of certain chlorinated compounds, e.g., chloroform, found in such waters will vary depending on the time of analysis unless the residual chlorine is consumed by a reducing agent such as sodium thiosulfate. Table II shows the concentrations of four compounds in Cincinnati tap water as a function of the sample age in days. The samples were taken from the distribution system at the EPA Environmental Research Center and maintained at 4°C until analyzed. No reducing agent was added to the samples.

The data in Table II show that under the experimental conditions used the chloroform concentration increased by 114% during the eight day storage period. Smaller increases in the concentrations of the brominated compounds were observed, but the carbontetrachloride concentration did not change within the precision of the method which averages approximately 6% in the 1-1000 ug/l range (15). A similar set of samples was stored at 22°C, and the same trends were observed except the concentrations of each of the halomethanes after seven and eight days were the same within

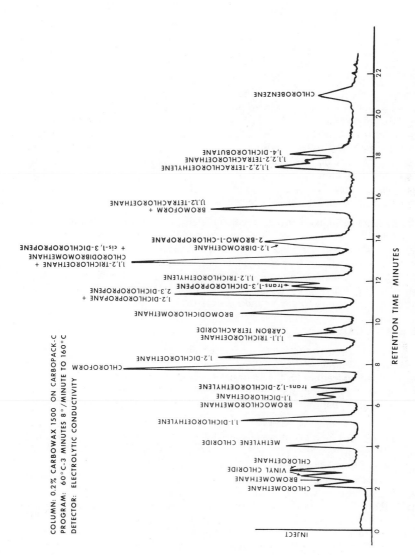

Figure 4. Chromatogram of organohalides

TABLE II. Effect of Residual Chlorine on Concentrations of
Chlorinated Methanes in Drinking Water at 4°C.

Time, days	CHCl$_3$ ug/l	CHBrCl$_2$ ug/l	CHBr$_2$Cl ug/l	CCl$_4$ ug/l
0	17.0	12.4	11.9	3.1
1	17.9	12.9	11.7	3.1
2	23.7	16.1	14.4	3.4
7	32.6	19.9	17.2	3.4
8	36.3	21.6	18.7	3.5

the precision of the method. This indicates that the rate of halogenation was reduced substantially after one week of storage at 22°C. The reduced rate of reaction was presumed due to a greatly diminished concentration of active halogenating agents, or the organic substrates, or both, but the details were not investigated further. In a third set of samples the reducing agent sodium thiosulfate was added at the time of sampling to quench the halogenation reaction. The concentration of the halomethanes in these remained constant within the precision of the method over the eight day period. In these samples no significant losses were observed indicating the effectiveness of the sample sealing procedures described in this paper.

The designer of an analytical survey that will include samples containing residual chlorine or other similar agents must be aware of these effects. The addition of a reducing agent at the time and place of sampling will give a measure of the instantaneous concentration of halogenated compounds, and this may or may not be the desired value. A measure of the maximum possible concentration of halogenated species in the particular sample may be obtained by storing the sample at 22°C or higher until the concentration of halogenated compounds is constant.

Selection of a GC Detector

The mass spectrometer GC detector has a high degree of compatibility with the purge and trap technique, and GC/MS has been employed widely with this isolation and concentration procedure for the analysis of volatile organic compounds. The mass spectrometer is strongly recommended for samples where there is a possibility of unexpected compounds, and for broad spectrum analyses [16] of poorly defined samples.

For well defined samples where the possibility of unexpected components is low, the flame ionization detector (FID) may be useful for the measurement of volatile hydrocarbons such as benzene, toluene, and the xylenes. However the FID is rendered somewhat unsuitable because the calibration procedure used with the purge and trap method requires the preparation of aqueous standards by mixing organic free water with aliquots of standard solutions in water miscible solvents such as methanol. Therefore the aqueous standards contain relatively large quantities of the miscible organic solvent relative to the compounds of interest. While these organic solvents are purged with relatively low efficiency, their relatively high concentration leads to significant interferences with the FID.

Another possible detector for well defined samples where the possibility of unexpected components is low is the [63]Ni electron capture detector (ECD). Application of this with the purge and trap detector is tedious because the ECD is so sensitive and susceptible to many trace interfering substances. Also temperature programming is usually desirable with the purge and trap procedure, and this is rather tedious with the ECD.

The halogen selective microcoulometric and electrolytic conductivity detectors have been applied extensively to the analysis of halogenated compounds in drinking water. Both have adequate sensitivity for the application and sufficient selectivity to allow reasonably accurate identifications with the retention time data. However, these detectors are also best applied to well defined samples where the probability of unexpected compounds is low.

Conclusion

The purge and trap method with GC/MS or a selective GC detector is an outstanding method for the analysis of volatile organic compounds in water samples. However the method requires careful calibration for quantitative analyses, and appropriate quality control procedures. Future work in analytical methods development is now turning to the other unsolved problems mentioned in the introduction. High performance liquid chromatography (HPLC) combined with mass spectrometry (LC/MS) or other selective detectors shows promise as a method for those compounds that are extracted into organic solvents, but are not amenable to GC because of insufficient vapor pressure or a tendency to decompose at elevated temperatures. Reverse phase HPLC shows promise as an approach to the analysis of polar water soluble organic compounds that are not partitioned efficiently into organic solvents.

Literature Cited

1. Harris, L. E., Budde, W. L. and Eichelberger, J. W., Anal. Chem. (1974), *46*, 1912.

2. Budde, W. L. and Eichelberger, J. W., J. Chromatogr. (1977), *134*, 147.

3. Richard, J. J. and Junk, G. A., J. Amer. Water Works Assoc. (1977), *69*, 62.

4. Henderson, J. E., Peyton, G. R. and Glaze, W. H., in "Identification and Analysis of Organic Pollutants in Water", L. W. Keith, Ed., 105-111, Ann Arbor Science, Ann Arbor, Mich., 1976.

5. "The Analysis of Trihalomethanes in Drinking Water by Liquid-Liquid Extraction", U.S. Environmental Protection Agency, Environmental Monitoring and Support Laboratory, Cincinnati, OH, 1977.

6. "EPA GC-MS Procedural Manual", Budde, W. L. and Eichelberger, J. W., Ed., U.S. EPA Report, in press.

7. Bellar, T. A. and Sigsby, J. E., EPA Report, Research Triangle Park, N. C., 1970.

8. Bellar, T. A. and Lichtenberg, J. J., J. Amer. Water Works Assoc. (1974), *66*, 739.

9. Bellar, T. A., Lichtenberg, J. J. and Eichelberger, J. W., Environ. Sci. Technol. (1976), *10*, 926.

10. Coleman, W. E., Lingg, R. D., Melton, R. G. and Kopfler, F. C., in "Identification and Analysis of Organic Pollutants in Water", L. W. Keith, Ed., 305-327, Ann Arbor Science, Ann Arbor, Mich., 1976.

11. Symons, J. M., Bellar, T. A., Carswell, J. K., DeMarco, J., Kropp, K. L., Robeck, G. G., Seeger, D. R., Slocum, C. J., Smith, B. L. and Stevens, A. A., J. Amer. Water Works Assoc. (1975), *67*, 634.

12. Brass, H. J., Feige, M. A., Halloran, T., Mello, J. W., Munch, D. and Thomas, R. F., in "Drinking Water Quality Enhancement Through Source Protection", R. B. Poja-sek, Ed., 393-416, Ann Arbor Science, Ann Arbor, Mich., 1977.

13. Kopfler, F. C., Melton, R. G., Lingg, R. D. and Coleman, W. E., in "Identification and Analysis of Organic Pollutants in Water", L. W. Keith, Ed., 87-104, Ann Arbor Science, Ann Arbor, Mich., 1976.

14. McAullife, C., Chem. Tech., (1971), 46.

15. Bellar, T. A. and Lichtenberg, J. J., presented at the ASTM Symposium on "The Measurement of Organic Pollutants in Water and Wastewater", Denver, CO., June, 1978.

16. Budde, W. L. and Eichelberger, J. W., Proceedings of the 4th Joint Conference on Sensing of Environmental Pollutants, New Orleans, Louisiana, November, 1977.

RECEIVED November 17, 1978.

Potentially Toxic Organic Compounds in Industrial Wastewaters and River Systems: Two Case Studies

RONALD A. HITES, G. A. JUNGCLAUS, V. LOPEZ-AVILA, and L. S. SHELDON

Department of Chemical Engineering, Massachusetts Institute of Technology, Cambridge, MA 02139

The identification and quantitation of potentially toxic substances in the environment requires the application of sophisticated analytical techniques. Ideally, these should exactly identify each of several hundred compounds present in very complex mixtures even though each species may have an environmental concentration of less than a part per billion. The most generally useful and widely employed analytical tool which meets these requirements is gas chromatography mass spectrometry (GCMS). In this paper, we will briefly review sample isolation methods which are used with GCMS and present two case studies on the organic compounds in industrial wastewaters and river systems which demonstrate these and other principles.

Concentration Methods. The GCMS analysis of an environmental sample starts with the isolation of the organic compounds from the matrix (air, water, food, etc.) into a form suitable for introduction into the GCMS instrument, typically a solution in a volatile solvent. This concentration step includes essentially three major methods: vapor stripping, solvent extraction, and lipophilic adsorption. We have recently reviewed the detailed operation of these methods (1), (See also Bellar, Budde and Eichelberger, this volume) but their general features will be outlined here.

In the vapor stripping technique, one bubbles an inert gas up through a column of water. This sparging effect strips the volatile organics from the water into the gas phase which is then passed through a cold trap or, more commonly, through a lipophilic trap such as Tenax. This trap is, in turn, placed into the gas chromatographic stream and heated. The organics are thus vaporized and analyzed by the GCMS system. The technique is most applicable to very volatile compounds such as chloroform, but by heating the water, compounds with volatilities up to the equivalent of a C_{20} normal hydrocarbon can be isolated. Vapor stripping is a very sensitive technique; part per trillion analyses of organic compounds in drinking water have been reported (2).

Solvent extraction is a very widely used and simple preconcentration technique. After the sample is extracted with a suitable solvent (such as methylene chloride), the extract is concentrated by evaporation and subjected to analysis. One important requirement is extremely clean solvents; fortunately these are now commercially available. Because of the evaporation step, solvent extraction cannot be used for the analysis of very volatile compounds. Depending on sample size, sensitivities of 0.1 ppb can easily be achieved.

In the lipophilic adsorption technique, large volumes of water are passed through

0-8412-0480-2/79/47-094-063$07.00/0

a column packed with a material which has an affinity for organic compounds; materials such as charcoal, XAD resins, and Tenax have been used. All of these adsorbents have a non-polar surface; thus, they tend to accumulate the lipophilic (fat soluble) compounds rather than more water soluble species. After sufficient water has passed through the column, the column is drained and extracted with a suitable solvent. The extract is then evaporated and analyzed by GCMS. For this reason, this technique has the same volatility limitations as solvent extraction. Lipophilic adsorption can, however, be used for the analysis of very large volumes of water (up to 500,000 liters; see reference 3), thus sensitivities are usually excellent.

These three concentration techniques have different applicabilities. They differ in their sensitivity, in the polarity of the analyte, and in their procedural difficulty. Obviously the proper technique must be selected based on the problem at hand. It may be wise to do an exploratory study with each technique to see what range of concentrations and compound types are present in a typical sample.

The remainder of this paper is a report on two case studies which will present the detailed organic analyses of two industrial wastewater and receiving water systems. In the first study, we will discuss the detailed environmental impact of a small specialty chemicals plant on its receiving water. In the second study, we will report on the detailed organic analysis of Delaware River water, a river which receives wastewater from many large-scale chemical producers. In these cases, we are concerned with the identities of compounds entering the receiving waters *and* sediments, the compounds already present, and others which may be formed through *in situ* transformations. Some of this information has appeared elsewhere (4).

Specialty Chemicals Plant

This plant operates in a batch production mode, generally following a weekly schedule. A wide range of compounds including pharmaceuticals, herbicides, antioxidants, thermal stabilizers, ultraviolet light absorbers, optical brighteners, and surfactants is produced. Water is used in synthetic processes, in solvent recovery, in steam jets, and in vacuum pump seals. The wastewater is neutralized in either of two one-million gallon equalization tanks, passed through a trickling filter for biological degradation, and clarified in a 150,000 gallon tank with a residence time of two hours. The water spills over from the clarifier at a rate averaging 1.3×10^6 gal/day and enters the river through an underground pipe about 100 yards away. Only about one fourth of the total BOD, which averages 12,000 lb/day, is removed by the waste treatment system and much of this is in the form of low molecular weight solvents.

Sampling. The water and sediment sampling sites in the vicinity of the plant are shown in Figure 1. Water samples were collected in one gallon amber glass bottles with Teflon-lined caps. Wastewater samples were collected as the water spilled over from the clarifier. River water samples were collected both upstream and downstream from the plant from bridges and from a small boat. About 300 ml of Nanograde (Mallinckrodt) methylene chloride and 15 ml of 12M hydrochloric acid were added to the water samples at the collection site (except those used for volatile organic analysis) in order to minimize biological degradation and to start the extraction.

Sediment samples were collected with a dredge-type sampler from a boat and also with the aid of a diver. One quart glass jars with aluminum foil-lined caps were used for the sediment samples; after collection they were placed in a box containing dry ice. The composition of the river bottom sediments varied from coarse sand ($>600\mu$) in the center of the river to coarse and fine silt toward the banks.

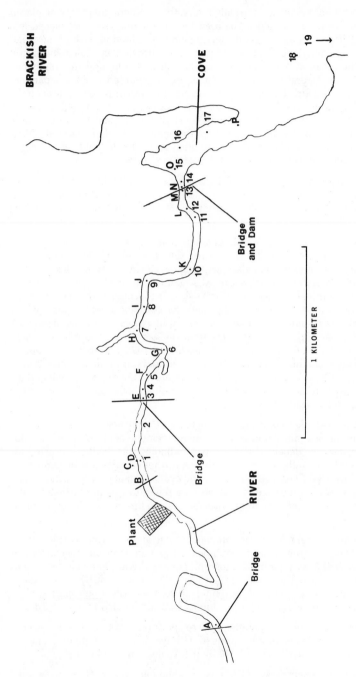

Figure 1. Map of the specialty chemicals plant and its environmental setting. Sampling sites are indicated: letters represent water samples and numbers represent sediment samples. Point C is the clarifier at the plant.

Procedures. When returned to the laboratory, the sediment samples were placed in a freezer and the water samples to be used for volatile analysis were placed in a refrigerator. A Teflon-covered magnetic stirring bar was added to each of the water samples containing methylene chloride for overnight extraction on a magnetic stirrer. Then most of the water was poured into another clean bottle, and the methylene chloride extract was separated from the remaining aqueous phase in a separatory funnel. A plug of pre-extracted glass wool was used to aid in phase separation for those samples containing emulsions. The extracts were rotary evaporated to the desired volume. It might be noted that Kaderna-Danish distillation may be more suitable for quantitative work. The decanted water from the acidic extraction was made alkaline with a pre-extracted, concentrated KOH solution and extracted with an additional 200 ml of methylene chloride to recover the basic compounds.

The exact concentrations of some of the volatile solvents in the wastewater were determined using direct aqueous injection of 2 µl aliquots onto a 2 m x 0.32 cm ID stainless steel column packed with 0.4% Carbowax 1500 on Carbopak C (Supelco, Inc.) and analyzed by GC/MS.

Qualitative analyses of the volatile organic compounds were performed using vapor stripping. About two liters of a river water sample or about 200 ml of a wastewater sample were put into a 3 liter glass stripping vessel similar to that described by Novotny et al. (5). The water temperature was maintained at about 80°C. Purified helium was passed through the sample from a glass frit located at the bottom of the apparatus at a rate of 120 ml/min. Helium and the stripped organics were passed through a water cooled condenser into two glass sampling tubes connected in parallel. These tubes were glass injection port liners from the gas chromatographs and were packed with about 40 mg of precleaned 60/80 mesh Tenax-GC porous polymer adsorbent. The liners were conditioned at 250°C for at least an hour in the injection port of the gas chromatograph prior to use. After vapor stripping for the desired length of time, the pre-columns were removed and stored in Teflon-lined screw-cap test tubes until analyzed by GC and GCMS.

The sediment samples were allowed to thaw at room temperature and then sieve-washed through a 2 mm stainless steel screen to remove pebbles and extraneous debris. Excess water was decanted and the wet sediment was Soxhlet extracted for several hours with Nanograde isopropyl alcohol. A further extraction with Nanograde benzene was necessary in order to isolate the polycyclic aromatic hydrocarbons. The isopropanol extract was evaporated to dryness on a rotary evaporator at 30-40°C; the benzene extract was freed of elemental sulfur by passage through a column of colloidal copper (6).

Instrumentation. Preliminary gas chromatographic analyses were carried out on a Perkin-Elmer 900 gas chromatograph equipped with a flame ionization detector and on a Hewlett-Packard 5730A gas chromatograph equipped with flame ionization and electron-capture detectors. The columns used were 180 cm x 2 mm ID glass columns packed with 3% SL-2100 (a methyl silicone fluid) on 80/100 mesh Supelcoport; we also used 25 m x 0.25 mm ID glass capillary columns statically coated (7) with SE-52.

Liquid chromatographic separations were performed on a Waters Model ALC/GPC 204 liquid chromatograph equipped with two model 6000 pumps, a model 660 solvent programmer, and a model 440 dual UV absorbance detector.

Low resolution (~800) mass spectra were obtained with a Hewlett-Packard

5982A GCMS system with a dual EI/CI source and interfaced to an HP 5933A data system. The quadrupole mass spectrometer was coupled to the gas chromatograph via a glass-lined jet separator held at 300°C. The mass spectrometer was usually operated in the electron impact mode, and spectra were obtained by continuous scanning under control of the data system. The instrument was also operated in the continuous scanning mode during analysis of collected liquid chromatographic fractions introduced with the direct insertion probe. Even when the LC fractions contained more than one compound, some fractionation of the compounds due to differential volatility allowed collection of relatively clean mass spectra for individual components.

High resolution (~20,000) mass spectrometric data were obtained on a Dupont 21-110B instrument with photographic plate detection. These plates were read on an automatic densitometer operated on line to an IBM 1800 computer (8).

Compound Identification. Our analyses resulted in the accumulation of thousands of spectra. These were initially compared to those published in the Eight Peak Index (9). Nearly all of the solvents, phenols, and hydrocarbons were found in this source, but the spectra of only a few of the remaining compounds were found here. Standard interpretation procedures were then used to identify the remaining compounds. Unfortunately, due to the normally high degree of aromaticity, the mass spectra of many of the unknowns consisted of only a few peaks, making identification difficult. However, a single high resolution mass spectral analysis of an extract provided the exact masses for most of the major molecular and fragment ions observed during a low resolution GCMS analysis. Thus, we were able to assign probable formulae to the various ions. This information often enabled us to identify structures related to other compounds already identified in the samples. When this was not successful, the Formula Index of Chemical Abstracts and the U.S. Trade Commission Report (10), which lists production data for individual and classes of compounds, were searched for related compounds produced by the company. The company advertising literature, which presented use and toxicity data for some of its products, was also useful in identifying a few compounds. Computer assisted library search routines (11,12) were not generally used. Because the Eight Peak Index contained most of the compounds in these data bases, time could be more economically utilized in performing the manual procedures described above.

Although the company officials and plant personnel cooperated fully in acquiring wastewater samples for analysis, they provided little information concerning identities of reactants of company products. Specific ring substitution patterns are not easily derived from mass spectrometric data and are included here only for cases where the literature specifies the isomer or in cases where we have purchased standards and observed identical mass spectra and gas chromatographic retention times.

Results and Discussion. All of the compounds identified in the wastewater, river water, and sediment samples are listed in Table I along with their concentration range. The individual concentrations have an estimated error of 20%. The structures of several of the compounds in Table I are given in Figure 2.

The identification of these 123 compounds (see Table I) was made possible only by the synergistic application of several analytical techniques. For example, the very high concentrations of a few compounds in most of the samples (e.g., no. 6, 10, 46, 81), precluded identification of many of the minor components during GCMS analysis. This dynamic range problem was solved, at least qualitatively, by HPLC followed by mass spectrometry.

Table I. Summary of All Compounds Found in Wastewater, River Water, and Sediment

Compound no.	Compound name	Concn range, ppm Wastewater	River water	Sediment	Present in tar ball [a]
	N-Containing heterocyclics				
1	Acetylpyridine	0.05[b]	ND	ND	No
2	Dibenzo[b,f]azepine*	<0.01	ND	ND	No
3	10,11-Dihydrodibenzo[b,f]azepine*	<0.01	ND	ND	No
4	5-(3-Dimethylaminopropyl)-10,11-dihydrodibenzo[b,f]azepine*	3.7[b]	ND	ND	No
5	4-*n*-Butyl-1,2-diphenylpyrazolidine-3,5-dione*	0.02–0.5	0.001–0.006	ND	No
6	2-(2'-Hydroxy-5'-methylphenyl)-2H-benzotriazole*	0.5–7	0.006–0.10	2–670	Yes
7	2-(Hydroxy-*t*-butylphenyl)-2H-benzotriazole*	ND	ND	60	Yes
8	2-(Hydroxy-di-*t*-butylphenyl)-2H-benzotriazole*	ND	ND	40	Yes
9	2-(Hydroxy-butyl-*t*-amylphenyl)-2H-benzotriazole*	d	<0.001	e	Yes
10	2-(2'-Hydroxy-3',5'-di-*t*-amylphenyl)-2H-benzotriazole*	0.55–4.7	0.007–0.085	1–100	Yes
11	2-(Hydroxy-*t*-butylphenyl)-chloro-2H-benzotriazole*	ND	ND	2–50	Yes
12	2-(2'-Hydroxy-3',5'-di-*t*-butylphenyl)-5-chloro-2H-benzotriazole*	ND	ND	2–300	Yes
13	2-Methoxy-4,6-bis-isopropylamino-*s*-triazine*	2–7.5	0.015–0.35	0.3[b]	No
14	2-Chloro-4,6-bis-isopropylamino-*s*-triazine*	0.5–3.5	0.01–0.45	4.5–12	No
15	Benzothiazole*	ND	0.002[b]	ND	No
	Nitrogen-containing compounds				
16	Aniline	0.02[b]	ND	ND	No
17	Acetanilide	0.2[b]	ND	ND	No
18	Di-*t*-butylcyanophenol	ND	ND	e	Yes
19	Azobenzene	0.03[b]	ND	ND	No
20	*N*-phenyl-1-naphthylamine	d	d	d	Yes
	Oxygen-containing compounds				
21	Methanol	17–80[h]	ND	ND	No
22	Stearyl alcohol	1–4	0.007–0.16	0.5	Yes
23	Acetone	200–230[h]	ND	ND	No
24	2-Butanone	8–20[h]	ND	ND	No
25	2-Methyl-3-heptanone	f	ND	ND	No
26	Phenol	0.01–0.30	0.01–0.10	ND	No
27	Cresol	0.07–0.15	0.001–0.01	ND	No
28	*t*-Butylphenol	0.001–0.15	0.003[b]	0.2–7	Yes
29	C$_8$-Alkylphenol (two isomers)	0.001–0.075	ND	5[b]	Yes
30a	2,6-Di-*t*-butylphenol	0.6–0.8	0.001–0.006	0.1–150	Yes
30b	2,4-Di-*t*-butylphenol	0.5–0.6	0.001–0.005	0.1–100	Yes
31	2,6-Di-*t*-butyl-4-methylphenol (BHT)	<0.01	0.001–0.002	1–60	Yes
32	2,4-Di-*t*-amylphenol	0.13–0.40	0.001–0.005	0.3–10	Yes
33	Tri-*t*-butylphenol	0.125[b]	ND	0.2–25	Yes
34	Methoxy-di-*t*-butylphenol	ND	ND	e	Yes
35	3,5-Di-*t*-butyl-4-hydroxybenzaldehyde	ND	ND	e	Yes
36	Nonylphenol	0.05[b]	ND	ND	No
37	Phenylphenol	ND	0.001–0.003	ND	No
38	Tetra-*t*-butyl-dihydroxydiphenyl*	ND	0.001–0.002	0.5[b]	Yes
39	Methylene-bis-(di-*t*-butylphenol)*	ND	ND	0.1–3	Yes
40	Ethylene-bis(di-*t*-butylphenol)*	ND	ND	e	Yes
41	Tetra-*t*-amyldihydroxydiphenyl*	ND	ND	d	Yes
42	Di-*t*-butylhydroxycinnamic acid*	d	d	d	No
43	Methyl-di-*t*-butylhydroxycinnamate*	d	0.010[b]	e	Yes
44	Ethyl-di-*t*-butylhydroxycinnamate*	ND	0.010[b]	e	Yes
45	3-(3',5'-Di-*t*-butyl-4'-hydroxyphenyl)propionic acid*	d	d	d	No
46	Methyl-3-(3',5'-di-*t*-butyl-4'-hydroxyphenyl)propionate*	0.6–11	0.025–0.20	1.5–170	Yes
47	Ethyl-3-(3',5'-di-*t*-butyl-4'-hydroxyphenyl)propionate*	<0.02	<0.002	2.5[b]	Yes
48	*i*-Propyl-3-(3',5'-di-*t*-butyl-4'-hydroxyphenyl)propionate*	0.02–0.30	<0.002	3.5[b]	Yes
49	Hexadecyl-3-(3',5'-di-*t*-butyl-4'-hydroxyphenyl)propionate*	0.040–1.4	0.002–0.04	ND	No
50	Octadecyl-3-(3',5'-di-*t*-butyl-4'-hydroxyphenyl)propionate*	1.6–6.0	0.008–0.2	8–220	Yes
51	C$_{22}$H$_{34}$O$_5$ (see footnote c)	0.06–0.9	0.006–0.012	0.6[b]	Yes
52	Di-(2-ethylhexyl)phthalate	ND	0.001–0.05	0.2–56	Yes
53	Di-octyl-phthalate	ND	0.001–0.02	1.5–25	Yes
54	Diphenyl ether	0.01–0.20	0.001–0.005	ND	No
55	Phenyl tolyl ether	ND	ND	e	Yes
56	C$_{29}$H$_{44}$O$_3$ (see footnote g)	0.06–0.20	<0.001	0.4–0.8	Yes
57	2,6-Di-*t*-butylbenzoquinone	0.01–0.02	0.001–0.011	0.1–40	Yes

Table I. Continued

Compound no.	Compound name	Wastewater	River water	Sediment	Present in tar ball[a]
			Concn range, ppm		
	Oxygen-containing compounds				
58	Tetra-*t*-butyldiphenoquinone*	0.02–0.30	<0.001	0.2–0.5	Yes
59	Tetra-*t*-butylstilbenequinone*	ND	ND	d	Yes
60	Binaphthylsulfone	ND	ND	d	Yes
	Halogenated compounds				
61	Dichloromethane	3–8[h]	f	ND	No
62	Chloroform	f	ND	ND	No
63	Trichloroethane	f	ND	ND	No
64	Trichloroethylene	f	f	ND	1no
65	Tetrachloroethylene	f	f	ND	No
66	Chlorobenzene	f	f	ND	No
67	Dichlorobenzene	f	f	e	Yes
68	Chlorotoluene	f	f	ND	No
69	Chlorobiphenyl	ND	ND	e	Yes
70	Chlorophenol	0.01–0.02	f	ND	No
71	Dichlorophenol	f	f	ND	No
72	Chloroaniline	ND	ND	1–2	Yes
73	Chlorotrifluoromethylaniline	ND	ND	0.2–180	Yes
74	Chlorophenylisocyanate*	ND	ND	0.1–2	Yes
75	Chlorotrifluoromethylphenylisocyanate*	ND	ND	0.1–13	Yes
76	4,4'-Dichlorocarbanilide*	ND	ND	d	Yes
77	4,4'-Dichloro-3-trifluoromethylcarbanilide*	ND	ND	d	Yes
78	4,4'-Dichloro-3,3'-bis-trifluoromethylcarbanilide*	ND	ND	d	Yes
79	Chlorotrifluoromethyldinitrobenzene	ND	0.002[b]	ND	No
80	Trichloro-amino-diphenylether	0.025–0.075	0.002–0.005	ND	Yes
81	2,4,4'-Trichloro-2'-hydroxydiphenylether*	6–14	0.012–0.30	1.2–5	No
82	Dichlorodibenzodioxin	d	d	ND	No
83	Trichlorodibenzofuran	d	d	ND	No
84	Bis-(dichlorophenoxychlorophenyl)-ether	d	ND	ND	No
	Aromatic hydrocarbons				
85	Benzene	f	f	ND	No
86	Toluene	13–20[h]	f	ND	No
87	Xylenes	f	f	ND	No
88	C$_3$-Alkyl-benzenes	f	f	30[b]	No
89	C$_4$-Alkyl-benzenes	f	f	e	Yes
90	C$_5$-Alkyl-benzenes	ND	ND	e	Yes
91	C$_6$-Alkyl-benzenes	ND	ND	e	Yes
92	C$_7$-Alkyl-benzenes	ND	ND	0.5[b]	Yes
93	C$_8$-Alkyl-benzenes	ND	ND	e	Yes
94	Biphenyl	ND	0.001–0.015	1–2	Yes
95	Terphenyl	ND	ND	e	Yes
96	Naphthalene	1–4	0.006–0.01	1.5[b]	No
97	Di-*t*-butylnaphthalene	ND	ND	e	Yes
98	Di-*t*-butyldihydroxynaphthalene	ND	ND	1[b]	Yes
99	Acenaphthylene	ND	ND	0.2–5	Yes
100	Fluorene	ND	ND	2–10	No
101	Methylacenaphthylene	ND	ND	5[b]	No
102	Methylfluorene	ND	ND	e	Yes
103	Hydroxyfluorene	ND	ND	0.1[b]	No
104	Phenanthrene	ND	ND	0.2–25	Yes
105	Methylphenanthrene	ND	ND	0.4–20	Yes
106	C$_2$-alkyl-phenanthrene	ND	ND	e	Yes
107	Fluoranthene	ND	ND	1–60	Yes
108	Pyrene	ND	<0.001	0.5–75	Yes
109	Methylpyrene (3-isomers)	ND	ND	0.2–15	Yes
110	Cyclopenta[c,d]pyrene	ND	ND	1–4	No
111	C$_{18}$H$_{12}$ PAH[i]	ND	ND	0.5–25	Yes
112	C$_{19}$H$_{14}$ PAH[i]	ND	ND	4–6	No
113	C$_{20}$H$_{12}$ PAH[i]	ND	ND	0.5–120	Yes
114	C$_{22}$H$_{12}$ PAH[i]	ND	ND	4–100	No
115	C$_{24}$H$_{14}$ PAH[i]	ND	ND	5–20	No

Table I. Continued

Compound no.	Compound name		Concn range, ppm			Present in tar ball [a]
			Wastewater	River water	Sediment	
		Alkanes				
116	C_6 alkanes		f	ND	ND	No
117	C_8 alkanes		f	f	ND	No
118	C_9 alkanes		f	f	ND	No
119	C_{10} alkanes		f	f	ND	No
120	C_{11} alkanes		f	f	73[b]	Yes
121	C_{12} alkanes		0.3[b]	f	ND	No
122	C_{13} alkanes		f f	ND	ND	No
123	C_{17} alkanes		0.1[b]	ND	ND	No

[a] The tar balls were small pea-size globules consisting of a complex mixture of anthropogenic compounds and were completely soluble in dichloromethane. They were found only at points 11 and 12 in the gravelly bottom just above the dam (Figure 1). Some of the tar balls were extracted along with the gravelly matrix material resulting in very concentrated samples, the quantitation of which would be meaningless. [b] Compound was detected in only one sample. [c] This compound has a molecular formula of $C_{22}H_{34}O_5$ as determined by high resolution mass spectrometry. Tentative identification is methyl isopropyl (3',5'-di-t-butyl-4'-hydroxy-benzyl)malonate. [d] The compound was isolated in a HPLC fraction and identified by mass spectrometry. Thus, quantitation was not possible. [e] The compound was only present in the tar ball-containing extract and thus could not be quantitated. [f] The compound was only identified through vapor stripping experiments and thus could not be quantitated. [g] This compound has molecular formula of $C_{24}H_{44}O_3$ from high resolution mass spectrometric data; a very tentative identification is dihydroxy-tetra-t-butyl-methyl-diphenyl ether. [h] Accurate compound quantitation based on direct aqueous injection. [i] A polycyclic aromatic hydrocarbon isomer. ND: Compound was not detected in the sample. [*] For structure, see Figure 2.

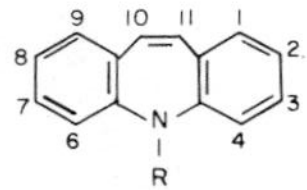

2 R = H
3 R = H, SATURATED AT $C_{10,11}$
4 R = $(CH_2)_3N(CH_3)_2$, SATURATED AT $C_{10,11}$

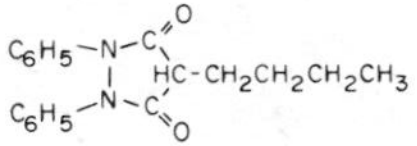

5

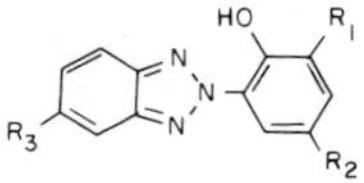

6 $R_1 = R_3 = H$, $R_2 = CH_3$
7 $R_1 = R_3 = H$, $R_2 = C(CH_3)_3$
8 $R_1 = R_2 = C(CH_3)_3$, $R_3 = H$
9 $R_1 = C_4H_9$, $R_2 = C(CH_3)_2C_2H_5$, $R_3 = H$
10 $R_1 = R_2 = C(CH_3)_2C_2H_5$, $R_3 = H$
11 $R_1 = H$, $R_2 = C(CH_3)_3$, $R_3 = Cl$
12 $R_1 = R_2 = C(CH_3)_3$, $R_3 = Cl$

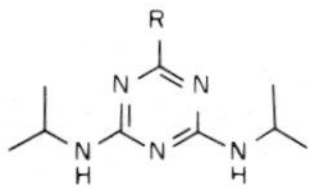

13 R = OCH_3
14 R = Cl

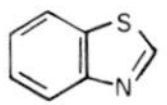

15

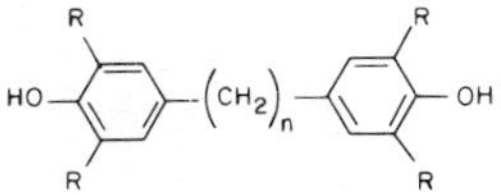

38 R = $C(CH_3)_3$, n = 0
39 R = $C(CH_3)_3$, n = 1
40 R = $C(CH_3)_3$, n = 2
41 R = $C(CH_3)_2C_2H_5$, n = 0

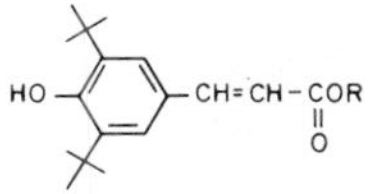

42 R = H
43 R = CH_3
44 R = C_2H_5

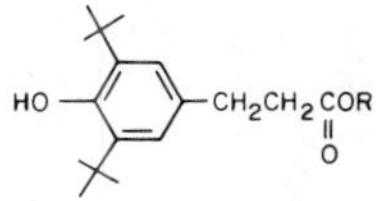

45 R = H
46 R = CH_3
47 R = C_2H_5
48 R = $CH(CH_3)_2$
49 R = $C_{16}H_{33}$
50 R = $C_{18}H_{37}$

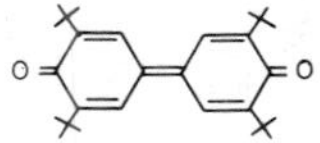

58

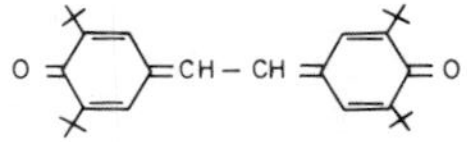

59

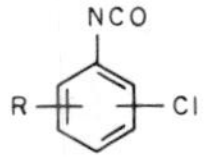

74 R = H
75 R = CF_3

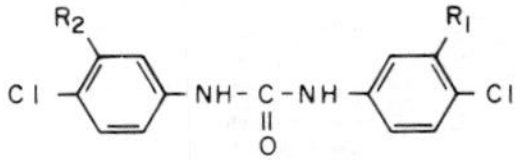

76 $R_1 = R_2 = H$
77 $R_1 = CF_3$, $R_2 = H$
78 $R_1 = R_2 = CF_3$

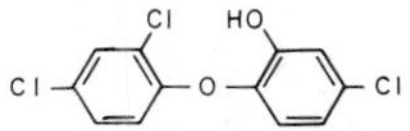

81

Figure 2. Structures of some of the compounds found in the wastewater of the specialty chemicals plant (see Table I)

Commonly Observed Compounds. Two of the substituted benzotriazoles (no. 6, 10 see Table I) were generally the most abundant anthropogenic compounds in the water and sediment samples. They are used as ultraviolet light absorbers in plastics and they possess antioxidant and thermal stabilization properties. No. 12 is apparently a former product which was found only in the sediments. The other benzotriazoles, present in much lower concentrations, are probably impurities in the major products. These benzotriazoles are characterized by resonance-stabilized internal hydrogen bonding of the phenolic hydroxyl to the benzotriazole ring; this apparently results in compounds with a high degree of environmental stability.

The phenols (no. 30a, 30b, 32) are used as reactants to synthesize several of the company's products, including the benzotriazoles. The 2-chloro-4,6-bis-isopropylamino-s-triazine (no. 13) are herbicides; the chloro compound is used to control weeds and grass in corn and in milo, and the methoxy compound is used for general plant control. Several esters of 3-(3',5'-di-t-butyl-4'-hydroxyphenyl) propionic acid were identified, the most abundant being the octadecyl and methyl esters.

Some ethers found in the wastewater and river water included diphenyl ether (no. 54), phenyltolyl ether (no. 55), and 2,4,4'-trichloro-2'-hydroxydiphenyl ether (no. 81) which is a backteriostat. The diphenyl ether became relatively more concentrated compared to other river water compounds in the brackish cove; this suggests high environmental persistence or its formation *in situ* from another compound. Some other very interesting compounds apparently resulting as by-products during the synthesis of 2,4,4'-trichloro-2'-hydroxydiphenyl ether, were trichlorodibenzofuran (no. 83), dichlorodibenzodioxin (no. 82), and a hexachlorinated compound of MW 558 (no. 84) which results from the condensation of two molecules of the ether with concomitant loss of water. We do not believe these compounds are artifacts produced from the ether during analysis because they were separated and collected as LC fractions, and the mass spectra were obtained with the direct insertion probe. These compounds have been detected in the wastewater, river water, and sediment.

Chlorotrifluoromethyl aniline (no. 73.) was found in the sediment samples. This compound is used as a reactant with chloro-aniline (no. 72) in the preparation of 4,4'-dichloro-3-(trifluoromethyl)-carbanilide, a disinfectant. Two other related compounds also found in some of the sediments were chlorophenyl isocyanate (no. 74) and chloro(-trifluoromethyl)phenyl isocyanate (no. 75). This suggests that some of the 4,4'-dichloro-3-(trifluoromethyl)-carbanilide may, in fact, exist in the sediment extracts but is decomposed in the injection port of the gas chromatograph, since it is very doubtful that the easily hydrolyzable isocyanates exist as such in the sediments. To strengthen this hypothesis some 3,4,4'-trichlorocarbanilide [none of the 4,4'-dichloro-3-(trifluoromethyl)-carbanilide was available] was analyzed by GCMS. The injection port temperature was 300°C. As expected, none of the parent compound eluted from the column. However, mass spectra were obtained for chlorophenyl isocyanate, dichlorophenyl isocyanate, chloroaniline, and dichloroaniline. The presence of the carbanilides themselves (no. 76, 77, 78) was confirmed with the help of HPLC and mass spectral identification.

Compounds present in the upstream river water extracts, but not in the plant wastewater included benzothiazole (no. 15), phenylphenol (no. 37), alkanes and phthalate esters (no. 52, 53). Their source is apparently another manufacturing plant.

Sedimentary Accumulation. The most striking result of this study is the high accumulation of anthropogenic compounds in the sediments. An estimation of the degree of sediment accumulation was gained by calculating a sediment accumulation factor, which is the average sediment concentration divided by the average river water concentration. Typical results are shown in Table II. The mechanism of this accumulation is not yet clear, but one possibility may be the gradual settling out of extremely fine particulate matter with a high organic load.

TABLE II. Sediment Accumulation Factors for
Some of the Anthropogenic Compounds from the Plant.
(The water and sediment concentration values are geometrical averages
of the high and low values given in Table I.)

Compound (see Fig. 2)	Wastewaster	River Water	Sediment	Accumulation Factor
6	2 ppm	0.02 ppm	40 ppm	2000
10	2	0.02	10	500
12	ND	ND	10	—
13	4	0.07	ND	0
14	2	0.07	7	100
30a	0.7	0.002	4	2000
30b	0.5	0.002	3	1500
31	ND	0.002	8	4000
38	ND	0.002	0.5	250
39	ND	ND	0.5	—
46	3	0.07	20	300
57	0.02	0.003	2	700
58	0.08	ND	0.3	—
80	0.04	0.003	ND	0
81	9	0.06	2	30

A small dam creates a falls (see Figure 1) where the river water empties into the brackish cove. Just above the dam, the sediment samples contained small tar balls (2-10 mm in diameter) consisting entirely of anthropogenic compounds; they were extracted along with some of the sediment. Quantitation of these samples would be meaningless, but they provided convenient samples for identifying large numbers of anthropogenic compounds in the sedimentary environment.

The sediment concentrations of anthropogenic compounds in the cove were somewhat less variable than upstream; this probably reflects the greater bottom uniformity of the cove. Fewer of the plant's compounds were detected in sediment from the channel where the cove leads into the brackish river (Point 18, Figure 1). Found at this location were various phenols (no. 28, 30a, 30b, 31, 33, 38, 39), di-t-butyl-benzoquinone (no. 57), 3,5-di-t-butyl-4-hydroxy-benzaldehyde (no. 35), three benzotriazoles (no. 6, 10, 12), 4,4'-dichloro-3(trifluoromethyl) carbanilide (no. 77), and 2-chloro-4,6-bis-isopropylamino-s-triazine (no. 14). The only compounds from the plant detected in the sediment sample from the brackish river (Point 19) were the two high molecular weight benzotriazoles (no. 10 and 12) and methyl 3-(3',5'-di-t-butyl-4'-hydroxphenyl) propionate (no. 46).

Chemical Transformation of Some of the Anthropogenic Compounds in the Environment. The fate of the di-*t*-butyl phenols is quite interesting. In the wastewater 2,6-di-*t*-butyl benzoquinone (no. 57) gives a very minor peak between 2,6-di-*t*-butyl-phenol (no. 30a) and 2,4-di-*t*-butylphenol (no. 30b). However, as the river water samples were taken further downstream from the plant, the relative intensity of the two phenol peaks gradually decreased and that of the benzoquinone increased. Two other quinones, tetra-*t*-butyl-diphenoquinone (no.58) and tetra-*t*-butyl-stilbenequi-none (no. 59), have also been identified. All of these quinones seem to be formed in the wastewater and river water by oxidation and coupling of the phenols *(13,14)*. However, in the reducing environment of the sediment, these quinones seem to be converted to the corresponding bisphenols. Another compound identified in the sediment extracts, probably resulting from oxidation of substituted phenols or substituted bis-phenols is 3,5-di-*t*-butyl-hydroxy-benzaldehyde (no. 35).

Delaware River Water

This section of this paper reports on a detailed study of the organic compounds in the Delaware River. Since most of our previous studies have dealt with isolated chemical plants on small rivers, we felt it would be instructive to study a complex river system surrounded by several large-scale chemical manufacturers. This would give us information on the effluents from several industrial dischargers, the relative persistence of chemical discharged from any single source, and the movement of these industrial chemicals in a major river ecosystem.

We selected the Delaware for several reasons: (a) Over 120 major chemical manufacturing plants (see Figure 3) are located along its banks and many discharge wastewater, either directly or indirectly, into the river *(15)*. (b) The Delaware is a major source of drinking water for many of the cities and counties in the area; for example, it provides 50% of Philadelphia's water *(16)*. (c) The incidence of cancer is very high in the areas surrounding the Delaware River. In fact, the highest bladder cancer rate in the United States occurs in Salem County, New Jersey *(17)*, an area of high chemical industry activity *(18)*. Based on these considerations, it is logical to ask if there is a correlation among cancer incidence, organics in the drinking water, and organics in the Delaware River. This is obviously a very complex and difficult question, but the answers must begin with a complete study of the organic compounds in the river itself.

The River System. The Delaware River is a 350 mile long waterway rising in central New York State, running through heavily industrialized areas of New Jersey, Pennsylvania, and Delaware to the Delaware Bay. Water flow in the lower third of the river from Trenton, New Jersey (river mile 132) to the Bay, is dominated by tidal action; tidal volumes around Philadelphia are at least an order of magnitude greater than the downstream river flow *(19)*. Thus, tidal action is responsible for a particularly slow downstream movement of the water. In general, any effluent discharged into the river will travel approximately 15 miles during a single tidal cycle with a net downstream movement of only one mile; total residence times in the river can be as long as 60 days *(19)*. Additionally, during periods of normal flow, the lower 55 miles of the river is an estuary characterized by salinity gradients; this limits both domestic and industrial usage *(20)*. Background data provided by the Delaware River Basin Commission show that the ratio of municipal to industrial discharge is approximately 2 to 1 *(15)*. Information provided on effluent source locations *(15)* aided in selecting representative sampling sites.

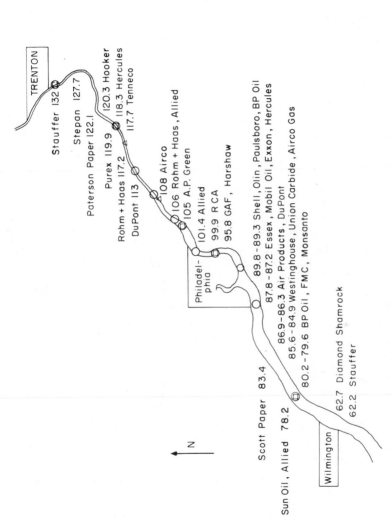

Figure 3. The Delaware River between river miles 60 and 140 showing locations of chemical companies (side not significant). Sampling sites: (○) collected August, 1976; (△) collected October, 1976; (□) collected March, 1977.

Sampling. In August 1976, 11 grab samples (3.5 1 each) suitable for solvent extraction were collected from the center channel of the Delaware River between Marcus Hook, Pennsylvania (river mile 78) and Trenton, New Jersey. In October, two additional samples (7.0 1) were taken from the shore for volatile analysis. Figure 3 gives water sampling locations. This particular river segment was of interest because it is the most heavily industrialized area along the river and is a direct source of drinking water for several of the cities in the region including Philadelphia. In addition, because of high tidal flows, this segment is well-mixed, but it is far enough upstream to avoid estuarine salinity effects.

Grab samples were collected in one gallon amber glass bottles with Teflon-lined caps at depths of 0.5 to 0.1 m. All center channel samples were collected aboard the *Aquadelphia*, the boat used by the City of Philadelphia for its own river sampling program.

Samples for volatile analysis were packed in ice to slow biological degradation prior to analysis, while samples for solvent extraction were immediately preserved by acidifying to pH 2 with hydrochloric acid and adding approximately 250 ml of methylene chloride. Addition of the organic solvent also started the extraction process. Sample workup was begun as soon as possible after returning to the laboratory, usually within 24 hours. In all cases, samples were kept refrigerated until analyzed. Analytical techniques for concentration, separation, and identification using solvent extraction, gas chromatography, and mass spectrometry have been discussed above.

The GC and GCMS analyses of the initial samples indicated low levels of all organics (in the sub-ppb range) and only gradual qualitative changes in the sample composition as a function of river location. Therefore, only five grab samples were collected in early March, 1977; these were larger in volume (21 1) and were more widely spaced than the first group. Their locations are given in Figure 3.

Sample Analysis. Initial results indicated that the samples were extremely complex, containing not only mixtures of industrial and natural organic compounds but also high background levels of gas chromatographically unresolvable materials. Organic background interferences were removed from the extracts using a silica gel chromatographic clean-up procedure. Methylene chloride extracts were evaporated to dryness and transferred to a column (5 x 0.6 cm ID) packed with deactivated silica gel (5% water). The sample was then fractionated by successively eluting with 10 ml each of hexane, benzene, and methanol. To remove fatty acid interferences, fractionated extracts were dissolved in methylene chloride and extracted with aqueous NaOH (pH 9-11). Blanks run for the concentration and clean-up steps indicated that no contamination was introduced during these procedures.

Identification of compounds in the river water extracts was based on the coincidence of gas chromatographic retention times and on the equivalence of electron impact and chemical ionization mass spectra with those of authentic compounds. Quantitation was based on standard curves generated for selected compounds.

Results and Discussion. The organic compounds identified in the Delaware River water samples are listed in Table III. These compounds cover a broad range of chemical types and include most of the major lipophilic compounds in the river. The data include the concentration range and the location of maximum concentration for each sampling season. Structures for a selected group of compounds are given in Figure 4.

Table III

Compounds Found in the Delaware River

Compounds	Vol.[a]	Winter Conc Range, ppb	Winter River Mile Max	Summer Conc Range, ppb	Summer River Mile Max
A. Isoprenoids					
1. 6,10,14-tri-methyl-2-penta-decanone*	–	ND[b]	–	0.8-2(11)[c]	106
2. α-terpineol*	–	0.5-4(5)[d]	98	ND[f]	–
3. chlorophyll[e]	–	ND	98	NQ[f]	106
B. Steroids					
4. cholesterol	–	5-10(5)	78	3-8(11)	93
5. cholestene	–	ND	78	trace	93
6. cholestanol	–	4-9(5)	78	1-2(11)	93
C. Fatty Acids and Esters					
7. stearic acid	–	NQ	98	NQ	–
8. palmitic acid	–	NQ	98	NQ	–
9. methyl stearate	–	ND	–	NQ	93
10. methyl palmitate	–	ND	–	NQ	93
11. methyl myristate	–	ND	–	NQ	93
D. Aromatic Hydrocarbons					
12. benzene	D	ND	–	ND	–
13. toluene	D	ND	–	ND	–
14. C_2 benzenes	D	17	98	ND	–
15. C_3 benzenes	D	4	98	ND	–
16. C_4 benzenes	D	trace	–	ND	–
17. C_5 benzenes	D	ND	–	ND	–
18. styrene	D	ND	–	ND	–
19. α-methyl-styrene	D	ND	–	ND	–
20. C_5 unsaturated benzene	D	ND	–	ND	–
21. naphthalene	D	0.7-0.9(3)	98	NA[g]	–
22. methyl naph-thalenes	D	0.4-1(3)	78,98	NA	–
23. C_2 naphthalenes	D	1-5(5)	78	NA	–
24. C_3 naphthalenes	D	2-5(5)	78	NA	–
25. C_5 naphthalenes	D	0.2-0.5(3)	78	NA	–

		Winter		Summer	
Compounds	Vol.[a]	Conc Range, ppb	River Mile Max	Conc Range, ppb	River Mile Max
26. pyrene	-	trace	-	NA	-
27. fluoranthene	-	trace	-	NA	-
28. anthracene	-	trace	-	NA	-
29. phenanthrene	-	trace	-	NA	-
30. methylphenan-threne	-	trace	-	NA	-
31. chrysene[h]	-	trace	-	NA	-
E. Phenols					
32. phenol	-	2-4(2)	98	ND	-
33. cresols	-	2	98	ND	-
34. C_2-phenols	-	2	98	ND	-
35. C_3-phenols	-	2	98	ND	-
36. C_4-phenols	-	2	98	ND	-
37. C_8-phenols	-	0.4-2(5)	105	trace	-
38. p(1,1,3,3-tetra-methylbutyl) phenol*	-	1-2(5)	98	0.2-2(11)	98
39. nonyl phenols	-	1-2(5)	105	0.04-1(11)	98
40. phenyl phenol	-	0.3	98	ND	-
41. cumyl phenol*	-	ND	-	trace	98
42. methyl isoeu-genol*	-	ND	-	NQ	-
F. Chlorinated Compounds					
43. chlorobenzene	D	7.0	98	ND	-
44. dichloroben-zene	D	0.4	98	ND	-
45. trichloroben-zenes	D	0.5-1(3)	98	ND	-
46. chlorotoluene	D	3	98	ND	-
47. benzylchloride	D	ND	-	ND	-
48. dichloromethane	D	ND	-	ND	-
49. chloroethylene	D	ND	-	ND	-
50. chloroform	D	ND	-	ND	-
51. trichloroethyl-ene	D	ND	-	ND	-
52. tetrachloro-ethylene	D	ND	-	ND	-
53. dichlorophenols	-	0.3	98	ND	-
54. trichlorophenols	-	2	98	ND	-

Compounds	Vol.[a]	Winter		Summer	
		Conc Range, ppb	River Mile Max	Conc Range, ppb	River Mile Max
55. chlorotrifluoro-methylaniline	-	trace-2(2)	78	ND	-
56. chlorotrifluoro-methylnitroben-zene	-	2-3(3)	78	ND	-
57. bis(chlorophenyl)ketone*	-	NQ	98	0.2-2(8)	93
58. bis(chlorophenyl)methanol*	-	NQ	-	0.1-1(8)	93
59. chlorophenyl-phenylmethanol*	-	NQ	98	trace	93
60. 1,1-bis(chloro-phenyl)-2,2-di-chloroethylene*	-	NQ	-	NA	-
61. chloromethyl acetophenone*	-	ND	-	trace	-
62. $C_{10}H_{11}Cl_3OS$[h]	-	trace	78	trace	78

G. Ethylene glycol derivatives

Compounds	Vol.[a]	Winter		Summer	
63. bis(2-chloro-ethyl)ether*	-	trace	98	ND	-
64. 1,2-bis(2-chloro-ethoxy)ethane*	-	15(2)	98	ND	-
65. 1-(2-chloro-ethoxy)-2-phen-oxy-ethane*	-	3	98	NQ	101
66. 1-chloro-2-[2(p-1',1',3',3'-tetramethylbutyl-phenoxy)ethoxy]ethane*	-	trace-2(3)	98	0.01-0.2(7)	101
67. 1-chloro-2-(2-[2-(p-1',1',3'-3'tetramethyl-butylphenoxy)ethoxy]ethoxy)ethane*	-	0.2-4(3)	98	0.03-0.1(7)	101
68. 2-(p-1',1',3',-3'-tetramethyl-butylphenoxy)ethanol*	-	NQ	-	NQ	-

		Winter		Summer	
Compounds	Vol.[a]	Conc Range, ppb	River Mile Max	Conc Range, ppb	River Mile Max
69. 2-[2-(p-1',-1,3',3'-tetra-methylbutyl-phenoxy)ethoxy]ethanol	–	NQ	–	NQ	–
70. 2-(2-[2-(p-1',-1',3',3'-tetra-methylbutyl-phenoxy)ethoxy]ethoxy)ethanol*	–	NQ	–	NQ	–
71. bis(n-butoxy-2-ethoxy-2-ethoxy)methane*	–	1-3(5)	78	2-3(11)	115

H. Esters (Plasticizers)

72. tri(t-butyl)phos-phate	–	0.4-2(4)	78	0.06-0.4(10)	88
73. tri(2-butoxy-ethyl)phosphate*	–	0.3-3(5)	78	0.4-2(11)	88
74. triphenylphos-phate	–	0.1-0.3(2)	78	0.1-0.4(11)	93
75. dibutyl phthal-ate	–	0.2-0.6(5)	–	0.1-0.4(11)	–
76. dioctyl phthal-ates	–	3-5(5)	–	0.06-2(11)	–
77. butylbenzyl-phthalates	–	0.4-1(5)	–	0.3-0.3(11)	–
78. dimethyl ter-phthalate	–	0.06	–	ND	–
79. di(2-ethyl-hexyl) adipate	–	0.8-0.3(5)	–	0.02-0.3(11)	–
80. di(isobutyl)azelate	–	NQ	98	ND	–
81. di(2-ethylhexyl)sebacate	–	trace	98	ND	–
82. tetraethylene-glycol-di(2-ethyl-hexanoate)*	–	1-14(5)	78	1-4(11)	106
83. tetraethylene-glycol-di(2-methyl-heptano-ate)*	–	0.1-0.3(5)	78	0.1-0.3(11)	106

		Winter		Summer	
Compounds	Vol.[a]	Conc Range, ppb	River Mile Max	Conc Range, ppb	River Mile Max
84. triethylene-glycol-di(2-ethylhexanoate)*	-	0.6-1(4)	78	ND	-
I. Others					
85. 2-ethylhexanol	-	3-5(2)	98	ND	-
86. 2,2,4-trimethyl-1,3-pentanediol-1-isobutyrate	-	1-6(5)	78	ND	-
87. 2,2,4-trimethyl-1,3-pentanediol-3-isobutyrate	-	1-4(5)	78	ND	-
88. 2-phenyl-2-propanol	-	2-3(3)	78	ND	-
89. isophorone*	-	trace	78	ND	-
90. nitroxylene	-	0.3	98	ND	-
91. o-phenyl anisole	-	trace	-	ND	-
92. binaphthyl-sulfones*	-	NQ	98	NQ	98
93. caffeine	-	trace	98	ND	-
94. methylcyclo-hexance	D	ND	-	ND	-
95. methylisobutyl-ketone	D	ND	-	ND	-
96. ethylthiopyri-dine*	-	ND	-	trace	103
97. phthallic acid	-	ND	-	NQ	-
98. 1,1,1-triphenyl-ethane	-	ND	-	NQ	-
99. fluorenone	-	NQ	-	NQ	-

Environmental Science & Technology

NOTES FOR TABLE III

a. <u>D</u> indicates compound was detected in the vapor stripping analysis of the October samples: quantitation was not possible.
b. <u>ND</u> indicates that the compound was not detected. c. Number indicates the number of samples out of 11 where the compound was found. When no number is shown the compound was detected in only 1 sample. d. Number indicates the number of samples out of 5 where the compound was found. e. Chlorophyll was identified from phytadienes (23) which are its volatile pyrolytic degradation products. f. <u>NQ</u> indicates that the compound was detected but for various reasons it was not quantitated. g. <u>NA</u> indicates that analysis for these compounds was not carried out. h. Exact structure not known. * For structure, see Figure 4.

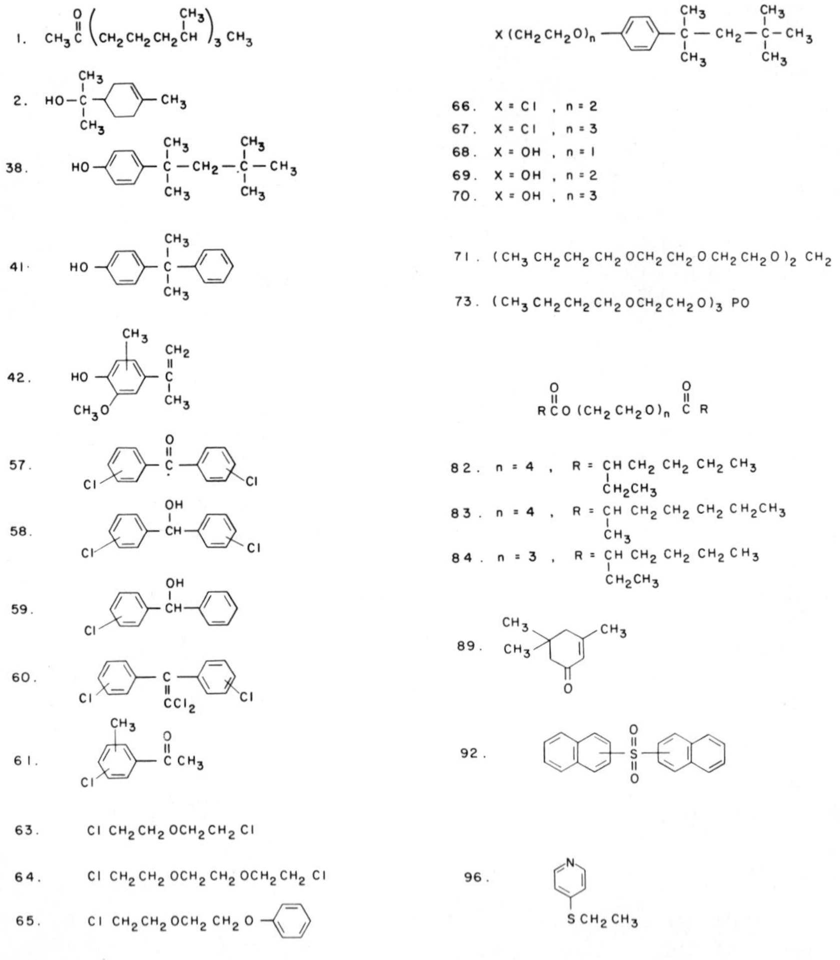

Figure 4. Structures of selected organic compounds found in the Delaware River (see Table III)

Since this study was designed to identify a broad spectrum of organic compounds, experimental procedures were as non-selective as possible. To prevent discriminating against any particular class of compounds, each sample was analyzed by GCMS prior to the clean-up procedures. All mass spectra found in the unfractionated extract were accounted for in the various fractionated samples.

The compounds listed in Table III are derived from three principle sources: natural products, municipal wastes, and industrial contaminants. Examples of each source are included in the following discussion.

Compounds 1, 2, and 3 are naturally occurring compounds resulting from the normal biological processes taking place in the river. The first compound, 6, 10, 14-trimethyl-2-pentadecanone, probably results from the oxidative degradation of phytol; other workers have found this ketone in various sediments (21) and plants (22). Chlorophyll was observed in our GCMS analyses as phytadienes which are produced in the injection port by pyrolysis of the phytol ester part of chlorophyll (23). Chlorophyll was abudant in the August water extracts, but it was a minor component in the winter water samples. This is not surprising since chlorophyll comes from algae and phytoplankton which are at a maximum concentration in the river during the summer months (16).

Municipal waste effluents are characterized by high concentrations of sterols, fatty acids, and fatty acid esters (24). These compounds (no. 4-11) were found at high levels in most of the samples from the Delaware River. For example, cholesterol was usually one of the most abundant compounds in the water. The concentration profile for cholesterol in the August water samples showed a maximum at river mile 93 which is consistent with the location of municipal sewage plants in the Philadelphia-Camden area. Fatty acids were not quantitated due to their poor chromatographic resolution, but they were present at very high levels in all samples.

The anthropogenic chemicals were by far the most numerous group of compounds and are the compounds of greatest concern to this study. In reviewing concentration and source data for these chemicals, it becomes apparent that they are of three types: Those found in industrialized urban areas with no specific production source; those commonly used in manufacturing processes with multiple sources; and those specific to a single industrial site and traceable to that source.

Included in the first group of general industrial contaminants are all of the aromatic hydrocarbons (no. 12-31, see Table III), most of the phenolic compounds (no. 32-42), most of the chlorinated species (no. 43-54), and some industrial solvents (no. 85,88,89,95). Almost all of these compounds have been isolated and identified in urban watersheds (2,25,26) where they appear to arise from automobile emissions, water chlorination, and general urban activities.

Source identification for several of the phenols, notable p-(1,1,3,3-tetramethylbutyl) phenol, and the nonylphenol isomers is more difficult. Concentration data shows highest levels around Philadelphia, implicating general urban activity as the primary source; however, there are several high production chemical companies in the area, one of which produces these phenols commercially (10). Under these circumstances no definite source can be identified.

Perhaps, the most ubiquitous of all environmental contaminants are the plasticizers (no. 72-84). These compounds can be found in the wastewater from a large number of industrial sources. The most common plasticizers (phthalates and adipates) show no concentration maxima along the river and may be assumed to enter from multiple locations. Some of the less common plasticizers tend to maximize in particular river segments suggesting single point sources. For example, tri (t-butyl) phosphate, and tri (2-butoxyethyl) phosphate, maximize near river mile 78. This concentration data is consistent with commercial production sites along the river (10).

The plasticizer, tetraethyleneglycol-di (2-ethylhexanoate), (no. 82), is interesting because it was both the most abundant compound in the river and the most challenging to identify. The electron impact mass spectrum for this compound (Figure 5) shows an intense peak at m/e 171 with less abundant ions at m/e 127, 99, 87, and 57. High resolution mass spectrometry estblished the elemental compositions of m/e 171 and 127 (see Figure 5). The small fragment ion at m/e 45 and the large neutral loss of 44 mass units (171 to 127) are characteristic of ethylene glycol compounds. The GC retention time suggested a rather high molecular weight despite the absence of any high mass fragments in the EI mode. Methane CI gave no additional information on molecular weight; but isobutane CI showed M+1 ion at 447. An elemental composition for the molecular ion of $C_{23}H_{46}O_7$ was hypothesized based on the rather saturated composition of m/e 171. Gas chromatography using a nitrogen-phosphorus detector did not contradict this hypotheses. A search of the EPA TSCA list (27) and Chemical Abstracts for industrial compounds corresponding to this molecular composition indicated that tetraethyleneglycol-di (2-ethylhexanoate) (a plasticizer patented and produced by one of the companies along the river) was a possibility. In fact, river water concentrations for this compound were highest in that sample taken adjacent to the suspected discharge site. Identification of the compound were verified using the authentic commercial product. The methyl-heptanoate isomer and the triethyleneglycol homolog (compounds 83 and 84, respectively) were identified in a similar manner.

The other polyethyleneglycol compounds (no. 63-71) are also industrial chemicals which are specific to a single source and which are traceable to that source. Identification of 1,2-bis (2-chloroethoxy) ethane in the river water near Philadelphia initiated a search for a possible source. According to the 1974 U.S. Tariff Commission Report (10), one of the companies in the area is the sole commercial producer of this compound, and holds a patent for its production (28). Similarly, 1-(2-chlorethoxy) 2-phenoxy-ethane, bis (2-chloroethyl) ether, and compounds 68, 69, and 70 are produced or patented by the same company.

Indentification of two other chloroethers (no. 66 and 67) was facilitated by their spectral and structural similarity to the above compounds. Figure 6 shows the EI mass spectrum of compound 67; the elemental composition of m/e 285 (obtained from HRMS) is included. An electron impact fragmentation pattern of m/e 63, 65, 107, 109, 151 and 153 is characteristic of a monochlorinated ion with 44 mass unit adducts. Previous identifications of chlorinated ethylene glycols suggested that this could be a similar compound with m/e 63 due to $ClCH_2CH_2$, m/e 107 to $ClCH_2CH_2OCH_2CH_2$, and m/e 151 to $ClCH_2CH_2OCH_2CH_2OCH_2CH_2$. Ions at 77, 91, and 135 are characteristic of C_3-phenolic compounds. A combination of these fragments accounts for the base peak at 285 (see Figure 6). A mass chromatogram indicated a very weak molecular ion at m/e 356 suggesting that a C_5H_{11} fragment should be added to the 285 ion to give compound 67. The hypothesized structure was synthesized by chlorinating the hydroxy compound (no. 70) with PCl_3 (28). The GC retention time and the mass spectrum for the unknown compound were identical to the synthetic compound. Compound 66 was similarly identified.

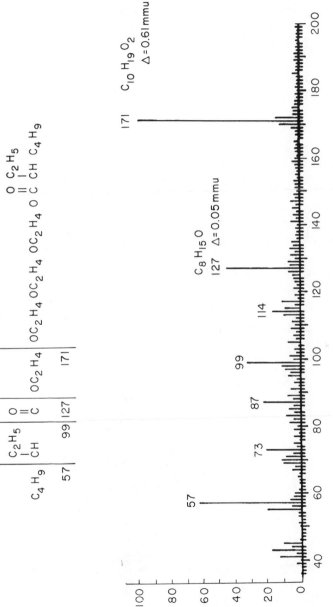

Figure 5. *Electron impact mass spectrum of tetraethyleneglycol-di(2-ethylhexanoate), compound 82. The elemental compositions of m/e 127 and 171 were established by high resolution mass spectrometry; the △ values indicate the error (in millimass units) between measured and calculated exact masses.*

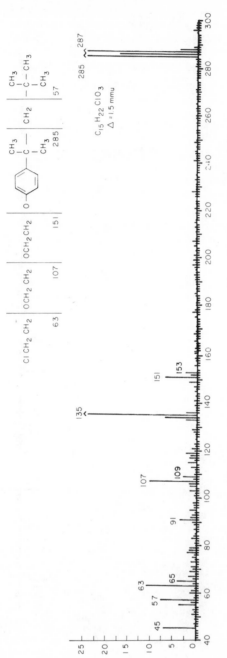

Figure 6. Electron impact mass spectrum of 1-chloro-2-(2-[2-(p-1′,1′,3′,3′-tetra-methylbutylphenoxy)ethoxy]ethane), compound 67. The relative intensities have been expanded by a factor of four; the intensities of the off-scale peaks are 135 (34%), 285 (100%), and 287 (37%).

Concentration profiles for compound 67 are given in Figure 7. Although the relative effect of dispersion due to tidal flow and downstream movement due to the net river flow are not precisely known, it is clear from these data that compound 67 comes from a point source located near river mile 100. In fact, the company which produces the precursor alcohols (no. 68-70), the chemically related chlorinated ethylene glycol (no. 64) and the C_8-phenol (no. 38) discharges its effluent at river mile 104.

The presence of these compounds in the Delaware River may have some health implications. If the discharge site at river mile 104 is correct, then these compounds would enter the river only four miles downstream from the inlet for Philadelphia's drinking water. Tidal action is sufficient to carry these chemicals upstream to the inlet and, in fact, the volatile ethers, bis-(2-chloroethyl) ether, and 1,2-bis(2-chloroethoxy) ethane, have been found in the drinking water supply *(29)*. Health effects, notably the carcinogenic activity, of these compounds are not known. It should be stressed that the higher molecular weight compounds (no. 65-70) have not yet been detected in the drinking water nor have their health effects been evaluated.

The chlorinated compounds (no. 57-60), bis (chlorophenyl) ketone, bis (chlorophenyl) methanol, chlorophenylphenylmethanol, and 1,1-bis (chlorophenyl)-2,2-dichloroethylene represent another important group of compounds traceable to a single industrial source. Although none of the compounds is manufactured commercially, the insecticide 1,1-bis (p-chlorophenyl)-2,2,2-trichloroethanol is produced by the same company which manufactures most of the ethyleneglycol compounds. We suggest that compounds 57 to 60 are either manufacturing by-products from the production of this compound or its environmental degradation products.

Finally, a few miscellaneous compounds which were identified in the Delaware River and which have not been previously reported as water contaminants will be discussed: Chloro (trifluoromethyl) aniline and chloro (trifluoromethyl) nitrobenzene (no. 55 and 56) were identified in the water, they had maximum concentrations at river mile 78. Both compounds represent common sub-structures in various pesticide and dye molecules, and several of the companies located along the river have patents using these compounds *(30-32)*. It is possible that these compounds are actually present in the river water as such, but it is also possible that they are formed in the GC injection port by pyrolytic degradation of larger pesticide or dye molecules (see above). All three binaphthyl-sulfone isomers (no. 92) were identified in the river water near Philadelphia. Product literature for one of the companies in the area indicates production of condensed sulfonated polymers derived from naphthalene sulfonic acid and maleic anhydride. It seems likely that the binaphthylsulfones are formed as by-products during preparation of this commercial product.

River water samples were collected both in August and early March; this allowed us to compare results from two sampling seasons. Generally, the two data sets were qualitatively similar, suggesting that pollution sources remained stable over the test period; however, two changes were observed. First, winter samples contained high levels of volatile organics (10-20 ppb) which were not detected in the summer water. Most likely high water temperatures (25-27° C) and turbulent river flow volatized organics from the river during the summer months.

The second change was the three to four fold increase in the level of almost all organics in the winter samples. This observation was corroborated by data on non-specific organic levels (COD, TOC, etc.) collected by another laboratory during the same sampling period *(16)*. Winter samples were collected during a period of high

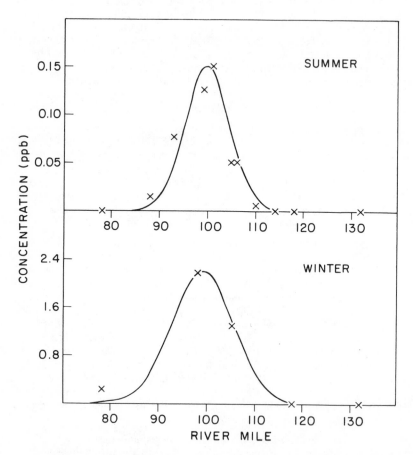

*Figure 7.   River water concentrations of compound 67 as a function of river mile
for samples collected in August, 1976 (summer) and March, 1977 (winter)*

water runoff and were very turbid in nature. It is possible that high levels of suspended particulate matter were responsible for increased organic concentrations in the water column. These particles could have been resuspending sedimentary organic compounds or they could have provided favorable adsorption sites within the water column for dissolved organics. As an alternative, municipal and industrial waste treatment systems may have been adversely affected by the cold winter temperatures; this would result in a significantly higher organic load entering the river.

Acknowledgement

This project was supported by the Research Applied to National Needs program of the National Science Foundation (grant number ENV-75-13069).

Literature Cited

1. Hites, R. A., Adv. in Chromatog. (1977) *15*, 69.

2. Grob, K., J. Chromatog. (1973) *84*, 255.

3. Rosen, A. A., Skeel, R. T., and Ettinger, M. B., J. Water Pollut. Contr. Fed. (1963) *35*, 777.

4. Jungclaus, G. A., Lopez-Avila, V., and Hites, R. A., Environ. Sci. Tech. (1978) *12*, 88.

5. Novotny, M., McConnel, M. L., Lee, M. L., and Farlow, R., Clinical Chemistry (1974) *20*, 1105.

6. Blumer, M., Anal. Chem. (1957) *29*, 1039.

7. Bouche, J. and Verzele, M., J. Gas Chromatog. (1968) *6*, 501.

8. Biemann, K., "Application of Computer Techniques in Chemical Research", pp. 5-19, Institute of Petroleum, London (1972).

9. "Eight Peak Index of Mass Spectra", Mass Spectrometry Data Center, AWRE, Reading, R67, 4PR, United Kingdom, (1974)

10. "Synthetic Organic Chemicals: U.S. Production and Sales, 1974", United States International Trade Commission, Publication No. 776, U.S. Government Printing Office, Washington, D.C. (1977).

11. Heller, S. R., Milne, G.W.A., and Feldmann, R. J., Science (1977) *195*, 253.

12. Hertz, H. S., Hites, R. A., and Biemann, K., Anal. Chem. (1971) *43*, 681.

13. Matsura, T., Yoshimura, N., Nischinaga, A., and Saito, I., Tetrahedron (1972) *28*, 4933.

14. Cook, C. D., J. Org. Chem. (1953) *18*, 261.

15. "Industrial Discharge Inventory," Delaware River Basin Commission, Trenton, N.J. (1975).

16. City of Philadelphia Water Department, personal communication (1977).

17. Hoover, R., Mason, T. J., McKay, R. W., and Fraumeni, Jr., J. F., Science (1975) *189*, 1005.

18. Hoover, R. and Fraumeni, Jr., J. F., Environ. Res. (1975) *9*, 196.

19. Harleman, D. R. F. and Lee, C. H., Technical Bulletin No. 16, Committee of Tidal Hydrology, Corps of Engineers, U.S. Army (1969).

20. Harleman, D. R. F. and Ippen, A. J., Proceedings A.S.C.E. (1969) *95*, 9.

21. Ikan, R., Baedecker, M. J., and Kaplan, I. R., Nature (1973) *244*, 154.

22. Kami, T., J. Agr. Food Chem. (1975) *23*, 795.

23. Hites, R. A., J. Org. Chem. (1974) *39*, 2634.

24. Garrison, A. W., Pope, J. D., and Allen, F. R., in Identification and Analysis of Organic Water Pollutants, L. H. Keith, ed., pp. 517-556, Ann Arbor Science Pub., Ann Arbor, MI (1976).

25. Shackelford, W. M. and Keith, L. H., "Frequency of Organic Compounds in Water," EPA-600/4-76-062, National Technical Information Service, Springfield, VA (1976).

26. Grob, K. and Grob, G., J. Chromatog. (1974) *90*, 303.

27. "Toxic Substance Control Act, PL 94-469, Candidate List of Chemical Substances," U.S. Environmental Protection Agency, Office of Toxic Substances, Washington, D.C. (1977).

28. Albright, R. L. and McKeever, C. H., U.S. Patent 3294847 (1966).

29. Suffet, I. H. and Radziul, J. V., J. Amer. Water Works Assoc. (1976) *68*, 520.

30. Theissen, R. J., German Patent 2261918 (1973).

31. Gerhard, W., German Patent 2203460 (1973).

32. Stoffel, P. J., U.S. Patent 3746762 (1973).

RECEIVED November 17, 1978.

Adsorbent Accumulation of Organic Pollutants for Bioassays

BONITA A. GLATZ

Food Technology Department, Iowa State University, Ames, IA 50011

COLIN D. CHRISWELL and GREGOR A. JUNK

Ames Laboratory, USDOE, Iowa State University, Ames, IA 50011

The attention given to organic compounds in drinking water has shifted from their role in causing undesirable odors and tastes to their potential effects on human health. These effects cannot be ascertained using data from traditional water quality parameters such as chemical oxygen demand [1], biological oxygen demand [2], total organic carbon content [3], fluorescence [4], ultraviolet absorbance [5] and carbon adsorbable material [6].

Some help for the evaluation of health effects has been provided by the progress made during the past decade in the identificaton of individual organic components present in drinking water. The number of identified compounds has grown to over 700 according to a recent compilation [7]. Bioassay results with some of these compounds have shown them to be toxic or potentially carcinogenic. However, all compounds identified in drinking water have not been tested thoroughly and some possibly detrimental substances have undoubtedly eluded even the most complete identification efforts.

An alternative to identifying organic chemicals in water and then determining their biological activity is the direct bioassay of mixtures accumulated from drinking water. This approach can provide part of the data for the preliminary assessment of health risks. The data can also be used to select those water sources on which the most strenuous identification efforts should be directed. Mixtures exhibiting considerable activity can be separated into chemical classes and bioassays can then be used to identify the active fractions. Ultimately the individual culprit chemicals would be identified. In this manner, efforts are directed towards the more detrimental components.

A forerunner to this approach was published in 1963 when organic chemicals accumulated from water by carbon adsorption and soxhlet extraction were found to cause tumors in laboratory animals [8]. However, the two-year period for completion of the animal tests of the accumulated mixture was a serious drawback.

Several short-term bioassay procedures [9-14] have been developed recently which are applicable to detecting mutagenic and potential carcinogenic activity of organic substances. The *Salmonella*/mammalian microsome assay or Ames Test [15-18] has been the most frequently applied and its efficacy has been well documented. This assay has also been applied to complex mixtures [19-22] to reduce greatly the time and effort required for a preliminary determination of adverse health effects. The *Salmonella* assay is rapid and inexpensive compared to whole animal testing and a close correlation has been demonstrated between mutagenic activity indicated by this procedure and carcinogenic activity [23, 24]. Preliminary results with mixtures isolated from drinking water have also been described [25, 26].

The organic material accumulated from water and used for bioassays should be representative of that originally present in the water. As much unaltered material as possible should be accumulated. In this sense, the desire for effective accumulation is the same whether the organic mixture is to be separated for identification purposes or used directly for bioassays. The primary intent of this report is to describe some accumulation techniques which are applicable to bioassay requirements and to present preliminary results of bioassays performed on organic mixtures accumulated from drinking water.

Accumulation Techniques

Accumulation as used in this report refers to the concentration of organic constituents within the water matrix or to their removal from water and their recovery in an unaltered form either in another matrix or as a neat mixture. Some traditional techniques for accumulating organic compounds from water include distillation, solvent extraction, freeze drying, and freeze concentration. Other more recent techniques include purge and trap [27] and closed loop purge and trap [28]. All these techniques are inconvenient and/or costly when large volumes of water must be processed to accumulate sufficient quantities of material for bioassay. These large volumes of water are better accommodated using reverse osmosis or solid sorbents such as activated carbon and a variety of synthetic polymers.

Reverse osmosis procedures concentrate over 90% of the total organic material present in water into an aqueous brine [29]. A problem has been the efficient transfer of the organic components to a solvent suitable for the bioassays [30]. Another problem is the loss of chemicals having molecular weights below 200-400.

Solid sorbents such as those identified in Table 1, have also been reported to be effective for accumulation of organic materials. The granular activated carbons have been most popular and they are normally used for removing organic impurities from drinking water and wastewater. They can also be used in analytical schemes for measuring organic contaminants. One of the first such procedures involved adsorbing contaminants on activated carbon, desorbing the organic compounds by soxhlet extraction, evaporating the solvent to dryness, and weighing the residue [31]. More definitive results can be obtained by applying gas chromatography-mass spectrometry techniques for separation and identification of the components in the residue. The Environmental Protection Agency used this approach during the National Organics Reconnaissance Survey [32].

The synthetic polymer sorbents are useful alternatives to the activated carbons for accumulating organic compounds [33-36]. In 1977 it was demonstrated that Amberlite XAD-2 resin was more efficient for accumulating many organic materials from water than was Filtrasorb 300 activated carbon (FS-300) [37]. In another study [38] greater amounts of gas chromatographable organic compounds were accumulated from water using the polystyrene-divinylbenzene copolymers, Amberlite XAD-2 and XAD-4 and Duolite L-863, than were accumulated using activated carbons, acrylic ester based resins, a carbonaceous resin, phenol-formaldehyde resins, and weak-base anion exchange resins. The composited plots of these data in Figure 1 illustrate clearly the

TABLE I. Identification of Solid Sorbents

Sorbent	Type	Supplier
XAD-2	Polystyrene-divinylbenzene resin	Rohm and Haas
L-863	Polystyrene-divinylbenzene resin	Diamond Shamrock
XAD-4	Polystyrene-divinylbenzene resin	Rohm and Haas
XAD-7	Acrylic ester resin	Rohm and Haas
XAD-8	Acrylic ester resin	Rohm and Haas
S-761	Phenol-formaldehyde resin	Diamond Shamrock
S-37	Anion exchange resin (weak-base)	Diamond Shamrock
A-7	Anion exchange resin (weak-base)	Diamond Shamrock
ES-561	Anion exchange resin (weak-base)	Diamond Shamrock
DARCO	Granular activated carbon	ICI-USA
FS-300	Granular activated carbon	Calgon
WVB	Granular Activated carbon	Westvaco
WVG	Granular activated carbon	Westvaco
G-216	Granular activated carbon	National Carbon
G-107	Granular activated carbon	National Carbon
XE-340	Carbonaceous resin	Rohm and Haas

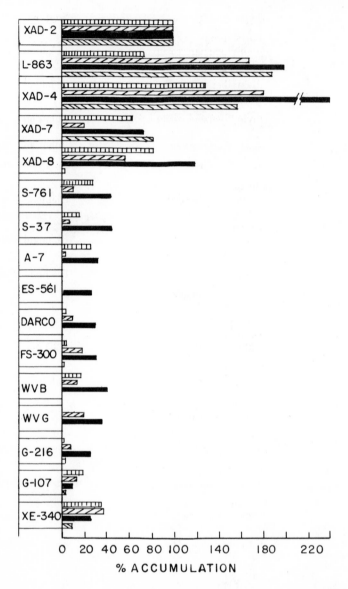

Figure 1. Comparison of gas chromatographable material accumulated by 16 solid sorbents at four water plants (see Table I for sorbent identification). Accumulation normalized to XAD-2.

superior accumulation by the styrenedivinylbenzene based resins. In the absence of any accumulation procedure which is 100% effective, the superiority of these resins suggests that they are a suitable, and probably the most desirable, starting point for accumulating organic components from water for bioassay purposes.

Experimental

One of the more efficient solid sorbents, Amberlite XAD-2 resin, has been used to accumulate organic materials from fourteen raw and finished waters [26].

Accumulations were performed in water utility facilities using 1/2" x 6" columns filled with the resin. Plant streams were sampled directly without any prefiltration or other treatment. Water was passed through the resin-filled columns at a flow rate of approximately 100 ml/min until a total volume of 200 L had been sampled. After sampling, each resin bed was eluted with 100 ml of diethyl ether, residual water dissolved in the ether was frozen out, and the ether was decanted into a concentration flask where the volume was reduced to 1.00 ml by distillation. Aliquots of samples accumulated at monthly intervals during the winter of 1976-77 were composited for use in performing bioassays. Dimethyl sulfoxide was added to each composite and the ether removed by free evaporation. The final volume of the dimethyl sulfoxide concentrates was adjusted so that each 10-μl aliquot would contain organic materials accumulated from 15 L of water. These concentrates were assayed for mutagenic activity using the spot test variation of the Ames test. Each sample was tested in replicate with each of the strains, TA98, TA100, TA1535, TA1537 and TA1538. In addition, assays were performed in duplicate with the same strains with the microsomal fraction of Aroclor 1254 activated rat liver added to each test plate. Each set of assays was accompanied by positive controls (known mutagens) and negative controls (solvent and sorbent blanks).

The bioassay results are given in Table II. Mutagenic activity was detected in eleven of the finished and six of the corresponding raw waters. These results indicate mutagenic agents can be accumulated from water using XAD-2 resin. Additional investigations have been performed to compare XAD-2 with other sorbents.

Organic contaminants were accumulated from the drinking waters of four central Iowa communities using sixteen different test sorbents. The accumulated materials were assayed for mutagenic activity and these results are listed in Table III. Three of the four waters showed activity with the most positive results obtained when the polystyrene-divinylbenzenes, XAD-2, L-863 and XAD-4, were used. Fewer positive results were obtained using the acrylic ester based resins, XAD-7 and XAD-8. Only marginal activity was observed when the phenol-formaldehyde polymer, S-761, was used. Use of all the other solid sorbents yielded either no or highly limited evidence of mutagenic activity.

Solid sorbent procedures are currently being compared to reverse osmosis techniques [25] for accumulating organic materials for bioassays but results are not yet available. The apparent and unexplained correlation between the accumulation of mutagenic and gas chromatographable material from water is also being investigated.

Conclusions

Polystyrene-divinylbenzene resins are the most efficient solid sorbents tested for the accumulation of mutagenic agents from water. This greater efficiency recommends their use in bioassay programs for the preliminary assessment of health risks associ-

TABLE II. Mutagenic Activity in the Organic Material
Accumulated from 14 Raw and 14 Finished
Water Supplies.[1]

Mutagenic Activity[2]

Water Supply	Without S-9		With S-9	
	Raw	Finished	Raw	Finished
1	−	+	−	0
2	−	+	−	+
3	0	+	+	+
4	0	0	−	−
5	−	+	−	0
6	−	−	−	−
7	−	0	−	−
8	0	0	0	0
9	−	−	−	−
10	+	+	+	+
11	+	+	0	+
12	−	+	−	+
13	−	−	−	−
14	+	+	+	+

[1]Sample size for each test adjusted to represent the accumulated components from 15 L of water. *Salmonella typhimurium* strains, TA98, TA100, TA1535, TA1537 and TA 1538 were used. Each sample was tested with and without the addition of microsomal fraction of activated rat liver designated S-9.

[2]Mutagenic activity was recorded if at least one strain showed an increased reversion frequency in response to the test sample. (−) = no activity detected; (+) = at least 2x the number of colonies appeared as on a control plate situated in a ring around the sample disc; (0) = an increased colony count was noted, but not 2x the control value, or not in a ring.

**TABLE III. Mutagenic Activity in Organic Material
Accumulted from 4 Water Supplies Using 16
Different Sorbents.[1]**

Sorbent	Mutagenic Activity-Site[2]			
	A	B	C	D
XAD-2	+	+	+	−
L-863	+	+	+	−
XAD-4	+	+	+	−
XAD-7	+	0	+	−
XAD-8	+	0	0	−
S-761	0	0	0	−
S-37	−	−	−	−
A-7	−	−	−	−
ES-561	−	−	−	−
DARCO	−	−	−	−
FS-300	0	−	−	−
WVB	−	−	−	−
WVG	−	−	−	−
G-216	−	−	−	−
G-107	−	−	−	−
XE-340	−	+	0	−

[1]Sample size for each test adjusted to represent the accumulated components from 10
L of finished water. *Salmonella typhimurium* strains, TA98, TA100, TA1535, TA1537
and TA1538, were used.

[2]See explanation of activity code in Table II footnote.

ated with ingestion of various waters. These sorbent procedures are readily upgraded to the accumulation of the larger amounts of material necessary for whole animal tests when these are deemed necessary. If the health assessments are followed by the chemical characterization of the organic material then specialized, more efficient accumulation procedures could be developed.

It is suspected that chlorination may play a role in the increased occurrence of activity for the finished water relative to the raw waters. This suspicion is based largely on the known production of trihalomethanes and the suspected production of other halogen containing compounds during chlorination. Any trihalomethanes accumulated by the resin procedure are largely lost during the subsequent distillation and evaporation steps, but remaining chlorinated products may be responsible for the observed mutagenic activity. However, treatment chemicals and treatment chemical impurities may not be ruled out as the responsible agents based on current results. Furthermore, treatment may have removed some substances that inhibit the mutagenic response of materials present in raw water. These explanations for the increased occurrence of mutagenic activity in finished water are clearly speculative and require more definitive studies for verification. No doubt, these studies will include the identification of the individual chemical(s) responsible for the mutagenic activity.

Acknowledgements

Primary fundings for these investigations were provided by the American Water Works Research Foundation and U.S. Environmental Protection Agency. The work was supported by the U.S. Department of Energy, Division of Environmental and Biomedical Research. The co-operation and advice of James S. Fritz, Harry J. Svec, Ron Webb, O. Thomas Love, Michael Taras and the various representatives of the fourteen water utilities are gratefully acknowledged.

Literature Cited

1. MOORE, W. A., KRONER, R. C. and RUCHHOFT, C. C. "Dichromate Reflux Method for Determination of Oxygen Consumed." *Anal. Chem.*, 21, 953 (1949).

2. SIERP, F. "A New Method for Determining BOD." *Ind. Eng. Chem.*, 20, 247 (1928).

3. DOBBS, R. A., WISE, R. H. and DEAN, R. B. "Measurement of Organic Carbon in Water Using the Hydrogen-Flame Ionization Detector." *Anal. Chem.*, 39, 1255 (1967).

4. SILVIA, A. E. "Detection and Measurement of Microorganics in Drinking Water." *J. N. Engl. Water Works Assoc.*, 87, 183 (1973).

5. DOBBS, R. A., WISE, R. H. and DEAN R. B. "The Use of Ultraviolet Absorbance for Monitoring the Total Carbon Content of Water and Wastewater." *Water Res.*, 6, 1173 (1972).

6. BUELOW, R. W., CARSWELL, J. K. and SYMONS, J. M. "An Improved Method for Determining Organics by Activated Carbon Adsorption and Solvent Extraction." *J. Am. Water Works Assoc.*, 65, 57 and 195 (1973).

7. McCABE, L. J. "Health Effects of Organics in Water Study." Presented at the American Water Works Association Water Quality Technology Conference, Kansas City, MO, Dec., 1977.

8. HUEPER, W. C. and PAYNE, W. W. "Carcinogenic Effects of Adsorbates of Raw and Finished Water Supplies." *Am. J. Clin. Pathol., 39,* 475 (1963).

9. CHU, E. H. Y. "Induction and Analysis of Gene Mutations in Mammalian Cells in Culture", in HOLLAENDER, A. (ed.), *Chemical Mutagens, Principles and Methods for Their Detection,* Vol. 2, p. 411, Plenum Press, New York, NY, 1971.

10. ABRAHAMSON, S. and LEWIS, E. B. "The Detection of Mutations in *Drosophila melanogaster"*, ibid., p. 461.

11. UNDERBRINK, A. G., SCHAIRER, L. A. and SPARROW, A. H. "Tradescantia Stamen Hairs: A Radiobiological Test System Applicable to Chemical Mutagenesis", ibid., Vol. 3, p. 171.

12. ZIMMERMAN, F. K. "Detection of Genetically Active Chemicals Using Various Yeast Systems", ibid., p. 209.

13. MOREAU, P. and DEVORET, R. "Potential Carcinogens Tested by Inductions and Mutageneisis of Prophage λ in *Escherichia coli* K12", in HIATT, H. H., WATSON, J. D. and WINSTEN, J. A. (eds.), *Origins of Human Cancer,* p. 1451, Cold Springs Harbor Laboratory Publ., Cold Springs Harbor, NY, 1977.

14. MISHRA, N. K. and DiMAYORCA, G. " In Vitro Malignant Transformation of Cells by Chemical Carcinogens." *Biochim. Biophys. Acta, 355,* 205 (1974).

15. AMES, B. N., DURSTON. W. E., YAMASAKI, E. and LEE, F. D. "Carcinogens are Mutagens: A Simple Test System Combining Liver Homogenates for Activation and Bacteria for Detection." *Proc. Nat. Acad. Sci. U.S.A., 70,* 2281 (1973).

16. AMES, B. N., LEE, F. D. and DURSTON, W. E. "An Improved Bacterial Test System for the Detection and Classification of Mutagens and Carcinogens." *ibid.,* 782 (1973).

17. McCANN, J. and AMES, B. N. "The *Salmonella*/Microsome Mutagenicity Test: Predictive Value for Animal Carcinogenicity", in HIATT, H. H., WATSON, J. D. and WINSTEN, J. A. (eds.), *Origins of Human Cancer,* p. 1431, Cold Springs Harbor Laboratory Publ., Cold Springs Harbor, NY, 1977.

18. AMES, B. N., McCANN, J. and YAMASAKI, E. "Methods for Detecting Carcinogens and Mutagens with the Salmonella/Mammalian-Microsome Mutagenicity Test." *Mutat. Res., 31,* 347 (1975).

19. EPLER, J. L., CLARK, B. R., HO, C. H., GUERIN, M. R. and RAO, T. K. "Short-term Bioassay of Complex Organic Mixtures. Part II, Mutagenicity Testing." Presented at the Symposium on Application of Short-term Bioassays in the Fractionation and Analysis of Complex Environmental Mixtures, Williamsburg, VA, Feb., 1978.

20. COMMONER, B., VITHAYATHIL, A. J. and DOLARA, P. "Mutagenic Analysis of Complex Samples of Air Particulates, Aqueous Effluents, and Foods." ibid.

21. FISHER, G. L. and CHRISP, C. E. "Physical and Biological Studies of Coal Fly Ash." ibid.

22. PELLIZZARI, E. D. and LITTLE, L. W. "Integrating Microbiological and Chemical Testing into the Screening of Air Samples for Potential Mutagenicity." ibid.

23. McCANN, J., CHOI, E., YAMASAKI, E. and AMES, B. N. "Detection of Carcinogens as Mutagens in the Salmonella/Microsome Test: Assay of 300 Chemicals." Proc. Nat. Acad. Sci. U.S.A., 72, 5135 (1975).

24. McCANN, J. and AMES, B. N. "Detectio;n of Carcinogens as Mutagens in the Salmonella/Microsome Test: Assay of 300 Chemicals, Discussion," ibid., 73, 950 (1976).

25. GLATZ, B. A., CHRISWELL, C. D., ARGUELLO, M. D., SVEC, H. J., FRITZ, J. S., GRIMM, S. M. and THOMSON, M. A. "Examination of Drinking Water for Mutagenic Activity." J. Am. Water Works Assoc., in press (1978).

26. LOPER, J. C. AND LANG, D. R. "Mutagenic, Carcinogenic and Toxic Effects of Residual Organics in Drinking Water." Presented at the Symposium on Application of Short-term Bioassays in the Fractionation and Analysis of Complex Environmental Mixtures, Williamsburg, VA, Feb., 1978.

27. BELLAR, T. A., LICHTENBERG, J. J. and KRONER, R. C. "The Occurrence of Organohalides in Chlorinated Drinking Water." J. Am. Water Works Assoc., 66, 703 (1974).

28. GROB, K. and ZÜRCHER, F. "Stripping of Trace Organic Substances From Water: Equipment and Procedures." J. Chromatogr., 117, 285 (1976).

29. KOPFLER, F. C. Personal Communications. Jan., 1978.

30. LANG, D. R. Personal Communications. Feb., 1978.

31. Standard Methods for the Examination of Water and Wastewater, 13th Edition, American Public Health Assoc., New York, NY, 1971.

32. KOPFLER, F. C., MELTON, R. G., LINGG, R. D. and COLEMAN, W. E. "GC-MS Determinations of Volatiles for (NORS) of Drinking Water", in KEITH, L. H. (ed.), Identification and Analysis of Organic Pollutants in Water, p. 87, Ann Arbor Science, Ann Arbor, MI, 1976.

33. BURNHAM, A. K., CALDER, G. V., FRITZ, J. S., JUNK, G. A., SVEC, H. J. and WILLIS, R. "Identification and Estimation of Neutral Organic Contaminants in Potable Water." Anal. Chem., 44, 139 (1972).

34. JUNK, G. A., RICHARD, J. J., GRIESER, M. D., WITIAK, D., WITIAK, J. L., ARGUELLO, M. D., VICK, R., SVEC, H. J., FRITZ, J. S. and CALDER, G. V. "Use of Macroreticular Resins in the Analysis of Water for Trace Organic Contaminants." J. Chromatogr., 99, 745 (1974).

35. RICHARD, J. J. and FRITZ, J. S. "Adsorption of Chlorinated Pesticides from River Water with XAD-2 Resin." Talanta, 21, 91 (1974).

36. BURNHAM, A. K., CALDER, G. V., FRITZ, J. S., JUNK, G. A., SVEC, H. J. and VICK, R. "Trace Organics in Water: Their Isolation and Identification." J. Am. Water Works Assoc., 65, 722 (1973).

37. CHRISWELL, C. D., ERICSON, R. L., JUNK, G. A., LEE, K. W., FRITZ, J. S. and SVEC, H. J. "Comparison of Macroreticular Resin and Activated Carbon as Sorbents." ibid., 69, 669 (1977).

38. CHRISWELL, C. D., FRITZ, J. S. and SVEC, H. J. "Evaluation of Sorbents as Organic Compound Accumulators." Presented at the American Water Works Association Water Quality Technology Conference, Kansas City, MO, Dec., 1977.

RECEIVED November 17, 1978.

Trace Metal Monitoring by Atomic Absorption Spectrometry

PETER BARRETT and THOMAS R. COPELAND

Department of Chemistry and the Institute of Chemical Analysis, Applications and Forensic Science, Northeastern University, Boston, MA 02115

Since the inception of atomic absorption spectrometric analysis (AAS) some 25 years ago it has become the single most widely used technique for the analysis of inorganic pollutants. The primary reasons for its extensive use are its simplicity of operation, speed, high sensitivity and versatility, relative freedom from interference and low cost. Well over 60 metals and metalloids are amenable to AAS analysis at concentration levels ranging from milligrams to femtograms per milliliter depending on the element, atomizer and conditions employed. The most significant disadvantage of AAS is the lack of simultaneous multi-element analysis capability. Virtually all analysis are for a single element at a time although various commercially available equipment permits analysis of two elements simultaneously or up to six elements sequentially under microprocessor control. Thus, AAS is most ideally suited for analyses where a limited number of trace metal concentrations are desired with high accuracy and precision. If a large number of elemental concentrations are to be determined, the advantages of a simultaneous multi-element technique such as plasma atomic emission are obvious.

Over 1000 articles dealing with atomic absorption are published each year making a truly comprehensive review a difficult task. What the authors have chosen to do is to discuss the recent advances in instrumentation and technique of particular interest to those persons involved in the analysis of air and water samples for trace metal toxins.

Basic texts discussing fundamentals of AAS as well as methodology have recently been published by Robinson (82), Pinta (76) and Schrenk (85). A more extensive and extremely useful text on atomic absorption and fluorescence is that by Kirkbright and Sargent (39). A wide variety of reviews of AAS are available, the most comprehensive being the "Annual Reports on Analytical Atomic Spectroscopy" (18). A useful categorized abstract compilation which covers conference proceedings as well as published papers is published bimonthly (7). A categorized bibliography is published yearly by Slavin and Lawrence (91). A general review of flame emission atomic absorption and atomic fluorescence which concentrates on advances in the techniques is published biennially in Analytical Chemistry (26).

Instrumentation

Recent advances in instrumentation have been primarily in improved atom reservoirs and automation. By far the most commonly used atom reservoir in AAS is the flame, either air acetylene (2400°K) or nitrous oxide acetylene (3200°K) for more

0-8412-0480-2/79/47-094-101$05.00/0
© 1979 American Chemical Society

refractory elements. In all conventional flame atom sources the sample is introduced into the flame by nebulizing the sample solution thereby producing a dispersion of small droplets. These droplets are then dried and the resulting solid decomposed, vaporized and reduced to free atoms by the flame. The efficiency of atom formation is then a complex function of numerous interrelated steps. The size of the nebulized droplets and droplet size distribution clearly affect desolvation rates. The exact composition of the dried solid particles can significantly affect decomposition rates. The electron density of the flame, flame composition, temperature and flame gas velocity (residence time) affect all processes involved. Given the number of processes required and the relatively short analyte residence time in the observation region of the most flames (on the order of 10ms) the overall efficiency of atom production is low.

A theoretical study of factors which affect pneumatic nebulizers (nebulizer geometry, capillary diameter, temperature fluctuations, gas pressure, solution viscosity, etc.) has been published by Heineman (24). Ultrasonic nebulization (which produces both smaller droplets and a narrower droplet size distribution) continues to attract attention (93, 95).

Electrothermal (non-flame) atomizers were developed to completely separate the desolvation, decomposition, and atomization steps. The longer time and variable temperature available for each step hopefully minimizes the effect of slow reaction rates and could therefore be expected to minimize matrix interference effects and maximize atom formation. Unfortunately matrix effects which are not as well understood as are flame interferences do exist with most electrothermal atomizers and are discussed below. Electrothermal atomizers have been constructed from Mo, Ta, W, Pt, and C. The vast majority are constructed of graphite due to its high electrical conductivity, good thermal conductivity and mechanical properties — it is easily machined and maintains structural integrity after repeated heated and cooling cycles. Sizes range from ~0.3cm O.D. carbon "cups" and "tubes" through tubes 0.3cm O.D. and ~3.5cm in length ("carbon furnaces") to furnaces 0.5cm in diameter and 20 to 30 cm in length (e.g. Woodriff and L'vov furnaces). The larger furnaces are maintained at constant temperature while smaller furnaces, cups, rods, and tubes are carried through drying, ashing, and atomization in several steps. For almost all commercially available atomizers a general mode of operation obtains — drying the sample near 100°C, ashing from 100 to 1500°C depending on the matrix and finally atomizing at 3000°C. All of the atomizers are sheathed with Ar or N_2 to minimize reaction with atmospheric O_2.

It is evident that with the discrete cycles of the non-flame atomizers several reactions (desolvation, decomposition, etc.) which occur "simultaneously" albeit over rather broad zones in a flame (due to droplet size distributions) are separated in time using a non-flame atomizer. This allows time and temperature optimization for each step and presumably improves atomization efficiencies. Unfortunately, the chemical composition and crystal size at the end of the dry cycle is matrix determined and only minimal control of the composition at the end of the ash cycle is possible, depending on the relative volatilities and reactivities of the matrix and analyte. These poorly controlled parameters can and do lead to changes in atomization efficiencies and hence to matrix interferences.

A comprehensive review of electrothermal atomization devices has been published (94). The review includes a discussion of commonly encountered problems such as atom loss through non-pyrolytic graphite, non-isothermal conditions, differences in peak height and peak area measurement, etc.

Electrothermal atomizers offer far greater sensitivity over flames and hence lowered detection limits. They also require significantly less sample — from 1 to 100μL per analysis as opposed to 3 to 5 mL per minute of flame analysis time. This clearly extends the technique in that analysis of micro samples is rather straightforward. However, electrothermal atomizers increase analysis time by at least a factor of four, increase atomizer cost by at least a factor of two, and decrease precision (primarily due to pipetting errors associated with small sample volumes). Since small sample volumes are used and the sensitivity of the method is quite high, the skill required in handling the sample (avoiding contamination, pipetting, preparation of stable dilute standards, etc.) is much higher than with flame analysis. In addition, most electrothermal atomizers are physically rather small placing more stringent requirements on optical alignment.

Sample Collection, Preparation, and Interferences

Inorganic trace analysis requires more than cursory attention to sample collection and treatment techniques since contamination is not only possible but often unavoidable. A good discussion of problems encountered in "extreme trace analysis" has been published by Tolg (99). The following section discusses some innovations in atmospheric and water sampling procedures.

Air

Atmospheric particulates ranging in size from 8μ to 0.01μ are most often collected on filter media. When more size-selective sampling is required various types of impactors are employed. The filters must be ashed, digested or leached for subsequent dissolution and AAS analysis. The filters commonly employed are paper, polymer membranes, fiberglass, glass, or graphite. Filter paper offers the advantages of being relatively free of trace metal contaminants and is easily wet or dry ashed. However, the pore size of paper filters and hence the particle diameter cut off is non-uniform and not well-known. Membrane filters have known pore sizes but it is often more difficult to solubilize the collected material. Manuals and bibliographies for their use have been published (61). Glass and fiberglass filters have inordinately high levels of metal contamination and acid pre-cleaning of the filters is often required. Carbon disks or cups have been employed as filtration devices. They are clean and efficient and since the collection device is also the atomizer, sample preparation is minimized. Woodriff and Lech (57) have used the graphite sample cups from his Woodriff furnace as air sampling filters. The cups are then simply inserted into the furnace for AAS analysis with no pretreatment or digestion required. The high sensitivity of this furnace and good collection efficiency of the graphite filters allow trace metal analysis on air samples of less than 100 ml. Matousek (62) has adapted this technique for the analysis of lead with a commercial AA spectrometer. Although in the above two studies the analyte was lead, this method can be utilized for other trace metals. This has been demonstrated by Begnoche and Risby (9). Hanssen et al (98) have compared the direct atomization technique with conventional leaching procedures and shown that increased precision is obtained with the direct method. A comparison of the efficiency of graphite and conventional filters was published by Skogerboe et al (90). Contamination problems encountered with graphite cups have been studied by Noller (69). Procedures to minimize contamination are presented.

The problem with filter sampling is that only particulate matter is collected. In many cases the concentration of organometallics is of interest. Robinson (83) has described a novel apparatus for the simultaneous collection of particulate and molecular lead.

For a more complete discussion of air sampling the reader is referred to compilations of official methods (29) book reviews (19), methods and bibliographies (84, 64) and specific methods found in the applications section.

Water

Procurement and preservation of water samples for AAS analysis are often not afforded the attention and effort required. The manner in which the sample is treated can greatly affect the analysis results. A complete discussion of water sampling is impossible herein, we have chosen to provide some guidelines and generalities. Further information can be obtained by consulting various reports and methods manuals (2, 65, 73). A review of the literature dealing with water pollution control which includes a section on analytical methods is published annually by the Water Pollution Control Federation (34). The biennial applications reviews in Analytical Chemistry includes a review of water analysis (17). Several other articles dealing with sampling and analysis are brought to the reader's attention (11, 27, 104, 92).

Ideally the measurement of trace analyte concentrations would take place *in situ* although this is seldom feasible. As an alternative the sample should be analyzed as soon as possible after collection. It is well established that trace inorganics are lost during sample storage either by adsorption onto container walls or incorporation into micro-organisms (bacteria, algae, etc). Minimization of this loss requires careful consideration of the type of container used, the decontamination procedure, and the preservatives (if any) added. Plastic, Teflon® and quartz containers are the least susceptible to adsorption. Moody and Lundstrom (66) have reviewed the selection and cleaning of various plastic containers. Preservatives commonly added to water samples include mineral acids, NaOH, and $HgCl_2$. Refrigeration or freezing of the sample also minimizes loss as well as chemical change.

The chemical or physical form of trace metals in water is often of interest. The form in which a specific element is present will often influence is toxic effects. For instance the chemical state of chromium affects its toxicity; i.e., Cr^{+6} is more carcino genic than Cr^{+3}. Kopp (48) has described the various forms in which metals may be present. The categories include dissolved metals, suspended metals, total metals, extractable metals and organometallics. In addition, Kopp describes sample preparation requirements for each category. Gibbs (20) has also studied metal species in river water. It should be obvious that the desired analytical result has to be considered beforehand. For example, if dissolved metal concentrations were desired and normal acid preservation performed, suspended metals could possibly be solubilized to a large extent. Both Hamilton (25) and Robertson (81) have shown vast differences between acidified and non-acidified samples. Many other publications have dealt with this subject (16, 37, 80, 30).

Finally, when ultratrace determinations are being performed it is often necessary to preconcentrate the sample or separate the analyte of interest from the matrix. The most commonly employed methods for preconcentration and separation of water samples include evaporation, chelation, coprecipitation, extraction, ion-exchange, chromatography, and electrochemistry. The procedure adopted will depend on the analyte, the form in which it exists, and the sample matrix.

Applications

The variety of "environmental" samples usually analyzed by AAS is large: ranging from the air and water samples discussed here to biological tissues, blood, urine, etc. Each of these different matrices present possibilities for interferences to occur which depend on the analyte, the matrix, sample treatment and the atomizer employed. All of the reviews mentioned above contain sections concerning interferences and procedures to minimize or eliminate them.

The most common interference is non-specific absorption or scattering. With either a flame or electrothermal atomizer any molecular species or fragments, "smoke" particles, or inorganic salt particles may either scatter or absorb light from the source. This will result in a decrease in light impinging on the photomultiplier and hence would be construed as an atomic absorption unless corrections are made. The correction is most often performed by "reanalyzing" the sample using a continuum light source. Due to the relatively wide spectral bandpass (0.1 to 1.0 nm) of the monochromator and the narrow line width of atomic lines (~0.003 nm) a very small fraction of the continuum source photons are of the proper frequency for atomic absorption. Thus, any drop in the continuum radiation reaching the detector is a measure of the scattering and molecular absorption in the vincinity of the atomic line and may be subtracted from the analytical signal. The measurements may be performed separately on a single beam instrument or by pulsing both lamps on and off alternately, a dual beam in time, single beam in space arrangement or by utilizing a magnetic field to Zeeman split the resonance atomic level and select absorbing and non-absorbing components with a polarizer. Certainly other arrangements are possible e.g., utilizing a non-absorbing line at a nearby wavelength but instrumental designs based on the above are most common in commercial instruments.

Other interferences which may occur in flame AAS are ionization of the analyte, formation of a thermally stable compound e.g., a refractory oxide or spectral overlap (very rare). Non-flame atomizers are subject to formation of refractory oxides or stable carbides, and to physical phenomena such as occlusion of the analyte in the matrix crystals. Depending on the atomizer size and shape, other phenomena such as gas phase reactions and dimerization have been reported.

Minimization of these interferences is usually accomplished by matching the matrix of standards and samples as closely as possible, using standard additions procedures or extracting the analyte from the matrix. Most extraction procedures employ dithiocarbamates and 4-methyl-2-pentanone (methyl isobutyl-ketone). Again, tables of specific procedures are found in many of the texts and reviews. In addition to liquid-liquid extractions, procedures involvng evolution of a gaseous form of the analyte are frequently used. Although only a limited number of elements are amenable to such analytical procedures (As, Se, Sb, Pb, Bi and Sn as hydrides; Hg and Ga as metals) the techniques are extremely sensitive. The atomizer employed may be a flame, a "cool" flame such as $Ar/H_2/$entrained air, electrically heated quartz or ceramic tube, or a simple absorption cell (for Hg). Reducing agents commonly used are Zn, $SnCl_2$ or $NaBH_4$ in acidic media. The gas evolution analysis offers two major advantages over solution analysis, the sample is presented to the atomizer as a single easily atomized species and scattering from the matrix is eliminated. The elimination of scattering is particularly helpful for elements such as As, Sb, and Se whose resonance lines lie in the 200nm region where scattering is more severe (the intensity of light scattered from small particles is inversely proportional to the fourth power of the wavelength). This

TABLE I. ANALYSIS OF WATER SAMPLES BY
ATOMIC ABSORPTION SPECTROMETRY

Element (s)	Sample Matrix	Automizer Type[1]	Ref.
Ag, Cd, Co, Cu, Fe Ni, Pb, Zn	EPA standard water	F	38
Ag, Cd, Pb, In	stream, rain water	NF	79
As	water	NF	97
As, Sb	surface ground water	F	35
As, Sb, Se	natural waters	NF	22
As, Se	water	F	75
Ca	water	F	103
Cd	seawater	NF	59
Cd	natural waters	F	100
Cd, Cu, Co, Fe, Mn, Ni, Zn	potable water	F	1
Cd, Cu, Co, Mn, Pb, U, Zn	snow, natural water, seawater	F	51
Cd, Cu, Cr, Mn, Ni, Pb, Zn	lake water	F, NF	96
Cd, Cu, Fe, Mn, Pb, Zn	river water	F	101
Cd, Cu, Pb	natural waters	F	53
Cd, Cu, Pb	natural waters	NF	63
Cd, Pb, Zn	sea water	NF	33
Co, Cr, Cu, Fe, Mn, Ni, Pb, Zn	natural water, seawater	F	28
Cu	seawater	NF	15
Cu, Fe	seawater	NF	55
Cu, Fe, Mn, Zn	natural waters	F	74
Cr	seawater	F	21
Cr	natural waters	F	72
Hg	river water	cold vapor	36
Hg	surface, sewage waters	cold vapor	13
Hg	seawater	NF	12
Hg	seawater	cold vapor	70
Mn	natural waters	NF	87
Mo	natural waters	F	50
Mo	seawater	NF	67
Pb	natural waters	F	52
Pb	potable	F	60
Se	water, industrial effluents	F	25
Se	wastewater	NF	8
V	seawater	NF	68
W	natural waters	F	54
Zn	natural waters	F	49

[1]F refers to flame atomic absorption spectrometry and NF to flameless atomic absorption spectrometry (e.g. carbon rod).

**TABLE II. ANALYSIS OF AIR SAMPLES BY
ATOMIC ABSORPTION SPECTROMETRY**

Element(s)	Sample Collection/Preparation	Atomizer Type[1]	Ref.
Ag, Be, Cd, Pb	C tube/direct analysis	NF	88
Al, Ca, Cd, Co, Cr, Fe, Mg, Mn, Ni, Pb, Zn	polymer filter/acid dissolution	NF	9
As	paper filter/ash/acid dissolution	NF	102
As, Se, Sb	membrane filters/acid digestion hydride	F	47
Ba, Be, Cd, Co, Cr, Cu, Ga, Mn, Mo, Pb, V, Zn	NBS fly ash/LiBO$_2$ fusion/acid dissolution	NF	71
Be, Cd, Co, Cu, Cr, Mn, Ni, Pb	filter/acid digestion	F, NF	19
Be	glass filter/acid digestion	F, NF	105
Cd, Co, Cr, Cu, Fe, Mn, Ni, Pb, Sb, V, Zn	impactor/ash/acid dissolution	NF	58
Cd, Cu, Mn, Pb	paper filters, acid dissolution	NF	98
Cd, Pb	filter/acid dissolution	NF	10
Cd, Pb	filter/acid dissolution	F, NF	14
Cu	glass, membrane filters/ash/ acid dissolution	F	40
Cr	PVC filter/extraction	F	45
Fe	glass, membrane filters/ash/ acid dissolution	F	41
Hg	Ag wool/heat	cold vapor	44
Hg	hopcalite/acid dissolution	cold vapor	78
Hg	carbon coated filter/acid dissolution	cold vapor	32
Hg	Au plate graphite/direct	NF	89
Mn	filter/acid dissolution	F	77
Mo	glass, membrane filters/ acid dissolution	F	56
Ni	filters, impingers, prec./ acid dissolution	F	3
Pb	filter/acid dissolution	F	31
Pb	electrostatic prec./acid dissolution	F, NF	5
Pb	graphite disk, carbon bed/direct	NF	83
Pb	glass filter/ash/acid dissolution	F	86
Pb	graphite cup/direct	NF	57
Pb	membrane filter/acid digestion	F	46
Ti	filters, impingers, prec./ acid dissolution	F	4
V	filters, impingers/acid dissolution	F	6
V	glass, membrane filters, ash/ acid dissolution	F	43
Zn	glass, membrane filters/acid dissolution	F	42

[1]F refers to flame atomic absorption spectrometry and NF to flameless atomic absorption spectrometry (e.g. carbon rod).

reduction in scattering coupled with the fact that all the analyte from a rather large sample may be reacted and swept into the atomizer in a short time affords great enhancement in sensitivity over conventional aspiration into a flame. However, chemical interferences such as presence of an excess of an easily reduced substance or kinetic problems such as different reduction rates for organic and inorganic analyte moieties may cause difficulties for some sample types.

Tables I and II list examples of analysis which have been performed by AAS on water and air samples. They are by no means comprehensive but were picked as examples of current methodology to acquaint the novice with the limits of detection and procedural difficulties commonly encountered.

References

1. Aldous, K.M., Mitchell, D.G., Jackson, K.W., Anal. Chem., 47, 1034 (1975).

2. Annual Book of ASTM Standards, Part 23, Water, Atmospheric Analysis, American Society for Testing Materials, 1916 Race St., Phil., PA 19103 (1973).

3. ANON, Amer. Ind. Hyg. Assoc. J., 36, 700 (1975).

4. ANON, Amer. Ind. Hyg. Assoc. J., 36, 707 (1975).

5. ANON, Amer. Ind. Hyg. Assoc. J., 36, 709 (1975).

6. ANON, Amer. Ind. Hyg. Assoc. J., 36, 711 (1975).

7. Atomic Absorption and Emission Spectrometry Abstracts, PRM Science and Technology Agency Ltd., London.

8. Baird, R.B., Pourain, S., Gabrielian, S.M., Anal. Chem., 44, 1887 (1972).

9. Begnoche, B.C., Risby, T.H., Anal. Chem., 47, 1041 (1975).

10. Billiet, J.B., Block, C., Demuynck, M., Janssens, M., Atmos. Environ., 9, 1099 (1975).

11. Ediger, R.D., At. Absorpt. Newsl. 12, 151 (1973).

12. Edwards, Lil., Oregioni, B., Anal. Chem., 47, 2315 (1975).

13. El-Awacly, A.A., Miller, R. B., Carter, M.J., Anal. Chem., 48, 110 (1976).

14. Eller, P.M., Haartz, J.C., Amer. Ind. Hyg. Assoc. J., 38, 116 (1977).

15. Fairless C., Bard, A.J., Anal. Chem., 45, 2289 (1973).

16. Feder, G.L., Ann. N.Y. Acad. Sci., 199, 118 (1972).

17. Fishman, M.J. Erdmann, D.E., Anal. Chem., 49, 139R (1977).

18. Fuller, C.W., Ed., "Annual Reports on Analytical Atomic Spectroscopy", 1976, Vol. 6, The Chemical Society, London, 1977.

19. Gallay, W., Environmental Pollutants — Selected Analytical Methods, Butterworths, London (1975).

20. Gibbs, R.J., Science 180, 73 (1973).

21. Gilbert, T.R., Clay, A.M., Anal. Chim. Acta, 67, 289 (1973).

22. Goulden, P.D., Brooksbank, P., Anal. Chem. 46, 1431 (1974).

23. Hamilton, E.I., Minski, M.J. Environ. Lett., *3*, 53 (1972).

24. Heineman, W., Fresenius' Z. Anal. Chem., *279*, 351 (1976).

25. Henn, E.L., Anal. Chem., *47*, 428 (1975).

26. Hieftje, G.M. and Copeland, T.R., Anal. Chem. 50, 300 R (1978).

27. Hinge, D.C., Chem. Ind. (London) 727 (1973).

28. Hiraide, M., Yoshida, Y., Mizuike, A., Anal. Chim. Acta. *81*, 185 (1976).

29. Intersoc. Committee, Methods of Air Sampling and Analysis, Amer. Pub. Hlth. Assn., Washington, D.C., 1972.

30. Issaq, H.J., Zielinski, W.L., Jr., Anal. Chem. *46*, 1329 (1974).

31. IUPAC Applied Chem. Division, Toxicology and Industrial Hygiene Sect., Pure Appl. Chem., *40*, 35.1 (1974).

32. Janssen, J.H., Van Deneik, J.E., Bult, R., Degroot, D.C., Anal. Chim. Acta, *84*, 319 (1976).

33. Jensen, F.O., Dolezal, J., Longmyhr, F.J., Anal. Chim. Acta *72*, 245 (1974).

34. J. Water Pollut. Control Fed., *48*, 998-1086 (1976).

35. Kan, K., Anal. Lett., *6*, 603 (1973).

36. Kiemeneij, A.M., Kloosterboer, J.G., Anal. Chem., *48*, 575 (1976).

37. King, W.G., Rodriguez, J.M., Wai, C.M., Anal. Chem., *46*, 771 (1974).

38. Kinrade, J.D., Van Loon, J.C., Anal. Chem., *46*, 1894 (1974).

39. Kirkbright, G.F. and Sargent, M., "Atomic Absorption and Fluorescence Spectroscopy", Academic Press, New York, 1974.

40. Kneip, T.J., Ajemain, R.S., Carlberg, J.R., Driscoll, J., Kornreich, L., Loveland, J.W., Moyers, J.L., Thompson, R.J., Health Lab. Scie., *10*, 226 (1973).

41. Kneip, T.J., Ajemain, R.S., Carlberg, J.R., Driscoll, J., Kornreich, L., Loveland, J.W., Moyers, J.L., Thompson, R.J., Health Lab. Sci., *11*, 117 (1974).

42. Kneip, T.J., Ajemain, R.S., Carlberg, J.R., Driscoll, J., Kornreich, L., Loveland, J.W., Moyers, J.L., Thompson, R.J. Health Lab. Sci. *11*, 134 (1974).

43. Kneip, T.J., Ajemain, R.S., Carlberg, J.R., Driscoll, J., Kornreich, L., Loveland, J.W., Moyers. J.L., Thompson, R.J., Health Lab. Sci., *11*, 240 (1974).

44. Kneip, T.J., Ajemain, R.S., Carlberg, J.R., Driscoll, J., Kornreich, L., Loveland, J.W., Moyers, J.L., Thompson, R.J., Health Lab Sci., *11*, 342 (1974).

45. Kneip, T.J., Ajemain, R.S., Carlberg, J.R., Driscoll, J., Kornreich, L., Loveland, J.W. Moyers, J.L., Thompson, R.J., Health Lab. Sci., *13*, 82 (1976).

46. Kneip, T.J., Ajemain, R.S., Carlberg, J.R., Driscoll, J., Kornreich, L., Loveland, J.W., Moyers, J.L., Thompson, R.J., Health Lab. Sci., *13*, 86 (1976).

47. Kneip, T.J., Ajemain, R.S., Grunder, F.I., Driscoll, J., Kornreich, L., Loveland, J.W., Moyers, J.L., Thompson, R.J., Health Lab. Sci., *14*, 53 (1977).

48. Kopp, J.R., Kroner, R.C., Trace Metals in Waters of the United States, U.S. Dept. of Interior FWPCA Div. of Pollution Surveillance, Cinn. OH (1967).

49. Korkisch, J., Goedl, L., Gross, H., Talanta, 22, 281 (1975).

50. Korkisch, J., Goedl, L., Gross, H., Talanta, 22, 669 (1975).

51. Korkisch, J., Sorio, A., Anal. Chim. Acta, 79, 207 (1975).

52. Korkisch, J., Sorio, A., Talanta, 22, 273 (1975).

53. Korkisch, J., Sorio, A., Anal. Chim. Acta, 76, 393 (1975).

54. Korvey, J.S., Goulden, P.D., At. Absorpt. Newsl., 14, 33.

55. Kremling, K., Petersen, H., Anal. Chim. Acta, 70, 35 (1974).

56. Kupel, R.E., Brauerman, M.M., Bryant, J.M., Bumsted, H.E., Carotti, A., Donaldson, H.M., Dubois, L., Welbon, W.W., Health Lab. Sci., 10, 218 (1973).

57. Lech, J.F., Siemer, D., Woodriff, R., Environ. Sci. Technol 8, 841 (1974).

58. Lee, R.E., Crist, H.L., Eriley, A., Macleod, K.E., Environ. Sci. Technol., 9, 643 (1975).

59. Lund, W., Larsen, B.V., Anal. Chim. Acta, 72, 57 (1974).

60. Maines, I.S., Aldous, K.M., Mitchell, D.G., Environ. Sci. Technol., 9, 549 (1975).

61. Manuals and bibliographies for use of membrane filters. Millipore Corporation, Bedford, MA. 01730.

62. Matousek, J.P., Brodie, K.G., Anal. Chem., 45, 1606 (1973).

63. Mesman, B.B., Thomas, T.C. Anal. Lett., 8, 449 (1975).

64. Methods for Chemical Analysis of Water and Wastewater, National Environmental Research Center, Cinn., OH 45268 (1971).

65. Methods for Examination of Water and Wastewater, 13th ed., American Public Health Assoc., 1015 18th St., N.W., Wash., D.C. 20036 (1971).

66. Moody, J.R., Lindstrom, R.M., Anal. Chem., 49, 2264 (1977).

67. Muzzarelli, R.A., Rocchetti, R., Anal. Chim. Acta, 64, 371 (1973).

68. Muzzarelli, R.A., Rochetti, R., Anal. Chim. Acta, 70, 283 (1974).

69. Noller, B.N., Bloom, H., Anal. Chem., 49, 346 (1977).

70. Olafsson, J., Anal. Chim. Acta, 68, 207 (1974).

71. Owens, J.W., Gladney, E.S., Atomic Absorp. Newsl., 15, 95 (1976).

72. Pankow, J.F., Janauer, G.E., Anal. Chim. Acta, 69, 97 (1974).

73. Parker, C.R., Water Analysis by Atomic Absorption Spectroscopy, Varian Techtron Pty Ltd., Springvale, Victoria, Australia (1972).

74. Pierce, F.D., Brown, H.D., Fraser, R.S., Appl. Spectrosc., 29, 489 (1975).

75. Pierce, F.D., Brown, H.R., Anal. Chem. 48, 693 (1976).

76. Pinta, M., "Atomic Absorption Spectrometry", Hilger, Ltd., London, 1975.

77. Rathje, A.O., Marcero, D.H., Amer. Ind. Hyg. Assoc. J., *36*, 713 (1975).

78. Rathje, A.O., Marcero, D.H., Amer. Ind. Hyg. Assoc. J., *35*, 571 (1974).

79. Rattonetti, A., Anal. Chem., *46*, 739 (1974).

80. Rosain, R.M., Wai, C.M., Anal. Chim. Acta., *65*, 279 (1973).

81. Robertson, D.E., Anal. Chim. Acta, *42*, 533 (1968).

82. Robinson, J.W., "Atomic Absorption Spectroscopy", 2nd ed., Marcel Dekker, New York, 1975.

83. Robinson, J.W., Rhodes, L., Wolcott, D.K., Anal. Chem. Acta., *78*, 474 (1975).

84. Saltzman, B.E., Burg, W.R., Anal. Chem. *49*, 1R (1977).

85. Schrenk, W.G., "Analytical Atomic Spectroscopy", Plenum Publishing Co., New York, 1975.

86. Scott, D.R., Hemphill, D.C., Holboke, L.E., Long, S.J., Loseke, W.A., Pranger, L.J., Thompson, R.J., Environ. Sci. Technol., *10*, 877 (1976).

87. Shigematsu, T., Matsui, M., Fujino, O., Kinoshita, K., Anal. Chim. Acta, *76*, 329 (1975).

88. Siemer, D., Woodriff, R., Spectrochim. Acta, *29B*, 269 (1974).

89. Siemer, D., Woodriff, R., Appl. Spec. *28*, 68 (1974).

90. Skogerboe, R.K., Dick, D.L., Lamothe, P.J., Atmosph. Environ., *11*, 243 (1977).

91. Slavin, S., Lawrence, D.M., At. Absorpt. Newsl., 16, 7, 89 (1977).

92. Spencer, D.W., Brewer, P.G., CRC, Crit. Rev. Solid State Sci., *1*, 401 (1970).

93. Stupar, J., Spectrochim Acta, Part B, *31*, 263 (1976).

94. Sturgeon, R.E., Anal. Chem., *49*, 1255A (1977).

95. Suddendorf, R.F., Gutzler, D.E., Denton, M.B., Spectrochim Acta, Part B, *31*, 281 (1976).

96. Surles, T., Tuschall, J.R., Jr., Collins T.T., Environ. Sci. Technol., *9*, 1073 (1975).

97. Tam, K.C., Environ, Sci. Technol., *8*, 734 (1974).

98. Thomassen, Y., Solberg R., Hanssen, J.E., Anal. Chem. Acta., *90*, 279 (1977).

99. Tolg, G. Talanta, *19*, 1489 (1972).

100. Topping, J.J., MacCrehan, W.A., Talanta, *21*, 1281 (1974).

101. Tweeten, T.N., Knoeck, J.W., Anal. chem., *48*, 64 (1976).

102. Walsh, P.R., Fasching, J.L., Duce, R.A., Anal. Chem., *48*, 1012 (1976).

103. Ward, D.A., Biechler, D.C., AU. Absorpt. Newsl., *14*, 29 (1975).

104. Whitby, F.J., Chem. Ind. (London) 88 (1974).

105. Zdrojewski, A., Dubois, L., Quickert, N., Sci. of Total Environ., *6*, 165 (1976).

RECEIVED November 17, 1978.

Determination of Trace Inorganic Toxic Substances by Inductively Coupled Plasma–Atomic Emission Spectroscopy

FRANK N. ABERCROMBIE

Montana Bureau of Mines and Geology, Montana College of Mineral Science and Technology, Butte, MT 59701

ROMANA B. CRUZ

Barringer Magenta, 304 Carlingview Drive, Rexdale, ONT M9W 5G2

As the industrial development for our society continues to produce new and different technologies, new analytical problems also occur — which challenge the environmental analytical laboratory staff. The laboratory may receive environmental samples for trace element analysis which encompass a variety of sample matrices including solid waste, waste water, drinking water, atmospheric particulates, particulates from emission sources, food and animal tissue. Although these problems may seem diverse with respect to either the source or the form of the sample, three common demands are expected of the laboratory. The analytical chemist is being challenged to analyze: i) for more variables ii) at greater sensitivity and iii) in pogressively increasing numbers of samples each year. The instrumentation that will satisfy these requirements needs to be capable of rapid, accurate, precise, sensitive and simultaneous multi-parameter analysis. The instrumentation should also be amenable to automation and capable of analyzing diverse matrices with a minimum of preparation, separation, and concentration steps. Inductively coupled argon plasma atomic emission spectroscopy (ICP-AES or ICP) allows a competent analytical chemist or spectroscopist to obtain, accurate and cost effective analysis of minor and trace-elements for almost any sample that can be dissolved in a suitable solvent to yield a solution or a slurry of finely divided particles. Early routine applications of the ICP occurred in the laboratories of Stan Greenfield[1] at Albright and Wilson and Velmer Fassel[2] at Iowa State University. Today, ICP is being used for trace elemental analysis in industrial and government laboratories throughout the world.

This chapter will describe the operation of an ICP and explain why certain physical parameters contribute to sensitivity and freedom from interferences. Commercially available, modular assembled (ICP-AES) systems will be discussed with respect to the general configurations which they employ. The origin of spectral interferences and their accomodation will be explained. The effect of operating parameters and data-processing requirements will be discussed. General as well as environmental applications will be enumerated and specific examples given.

Operating Principles — There are many similarities between ICP-AES and the combustion flame spectroscopy techniques of flame atomic emission (FAE) and flame atomic absorption (FAA). In fact, the source of the ICP-AES has been referred to by Fassel as an "electric flame." The final prepared analytical sample is presented in liquid form for analysis except for unique situations. The liquid sample is drawn (or pumped) into a nebulizer to form an aerosol which is swept into the "flame" in a manner analogous to conventional FAA analysis. With ICP-AES, the aerosol is swept into the plasma by an argon carrier gas and evaporated by the plasma at the extreme temperatures, leaving salt particles. The particles are dissociated and excited by the plasma environment into atoms or ions, emitting the characteristic emission spectra of the elements present. This emitted radiation is focused upon the entrance slit (or grating) of a monochromator (or polychromator) in a manner similar to FAE and other emission spectroscopy techniques. The grating disperses the radiation into discrete wavelengths which pass through the receiver of exit slit(s) and are monitored by photomultiplier(s). After appropriate amplification of the photomultiplier current(s), the signal(s) may be monitored as current or voltage signal(s). For single-element ICP-AES, the signal may be read out on a meter or recorder. Multi-channel photo multiplier signals, either after or before amplification, are generally aquired and processed by a computer.

The instrument is calibrated by solution standards, which contain the elements of interest in an appropriate matrix. The concentration range covered for ICP-AES may be several orders of magnitude. Standards, blanks, and samples are analyzed in a sequence appropriate with the instrumental stability and precision desired. During nebulization, approximately fifteen seconds is required to obtain a steady-state signal; another ten seconds is required to integrate the signal; and a thirty second blank rinse is required to clean out the spray chamber. The actual time intervals will vary from system to system.

In order to appreciate the many advantages of ICP, it is extremely helpful to have a basic understanding for the operating principles of the plasma. Several reviews (4-9) are available describing plasma source principles so that only a brief discussion will be given here.

A plasma possesses all the properties of a gas but also contains sufficient positive ions and electrons needed to conduct electric current and to interact with magnetic fields. The electrons in the source are excited by a radio frequency (r.f.) magnetic field, created by oscillating currents in a induction (load) coil surrounding the plasma. The load coil is analogous to the primary windings of a simple transformer and the plasma as a one-turn secondary winding. In this manner, energy is inductively coupled into the plasma as the alternating magnetic field from the load coil induces electric eddy currents in the plasma. The resistance to the induced circulating electron current results in Joule heating. (Similar eddy currents will be induced in any conductor such as an iron or carbon rod placed inside the load coil.) The accelerated electrons sustain the plasma through collisional ionization of argon atoms which reach a steady-state equilibrium with ion recombination processes. During ignition the electrons are seeded into the argon gas stream either by sparking a tesla coil or by heating a conducting rod (carbon) to incandescence and removing it after the plasma strikes(10).

The plasma is thermally and electrically insulated from the load coil by the walls (usually quartz) of the plasma torch. The walls are cooled, to prevent melting, by a continual flow of gas (generally argon). Flow rates for analyzing agueous solutions are approximately 10 to 14 l/min. for a 1600 watt plasma torch, with the configuration depicted in Figure 1. Flow rates for analysis of organic solutions are usually higher.

The constricted orifice of the concentric quartz wall directs the coolant argon along the outer wall of the torch (Figure 1). The low-pressure region, generated by the Bernoulle effect from the peripheral high-velocity gas flows, centers the plasma, caus-

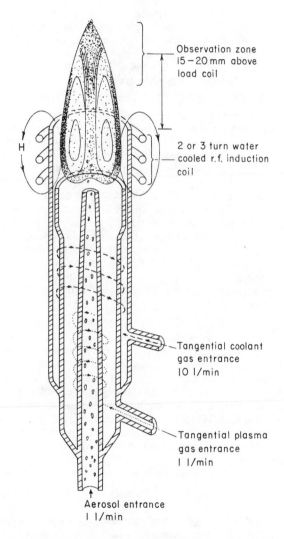

Observation zone
15–20 mm above
load coil

2 or 3 turn water
cooled r.f. induction
coil

H

Tangential coolant
gas entrance
10 l/min

Tangential plasma
gas entrance
1 l/min

Aerosol entrance
1 l/min

Figure 1.   Cross section of induction-coupled plasma torch, plasma, and radio
frequency coil

ing it to "float" within the torch. Although the coolant flow exerts a vector in the direction of flow upon the plasma, the magnetic fields maintain the plasma within the load coil, preventing the plasma from being forced from the torch by the flowing argon stream. This can be experimentally verified by the extremely hazardous operation of sliding a plasma torch containing a plasma back and forth along the magnetic field axis. Varying the r.f. power will also cause the plasma to move along the axis to a lesser degree as the force of the flowing gases counteracts the magnetic field. A flow of approximately 1 l/min. is used to cool both inner quartz wall edges, reducing the devitrification rate to the extent that a "precision" torch will last at least six months. The plasma gas flow can be increased slightly for organic solvents to prevent deposition of pyrolytic carbon on the inner quartz wall edges.

The innermost quartz channel is the injector channel, which carries the sample aerosol argon stream (carrier gas) to the plasma. Greenfield[1] recognized the importance of the injector design for directing the carrier gas stream through the plasma instead of around the plasmas peripheral edge. The injected plasma is a toroid created by the carrier gas passing between the high temperature (9,000 to 10,000° K) zones of the toroid lobes in which the main conduction current flows. The temperature in the carrier channel measured by Mermet[11] is reported to be about 6,000° K. This temperature, in a rare gas atmosphere, contains energetic ions and favors complete dissociation of all refractory materials studied to date. It prevents the occurrence of common solute matrix vaporization interferences that are usually encountered with most other atom reservoirs. If the aerosol does not pass through the central channel, solute matrix interferences are observed[12, 13]. This may well be the origin of many other interference examples reported prior to the early 70's. Solute matrix interferences will also be observed if the residence time of the aerosol within the central channel is decreased from the optimum[14].

Instrumentation — Many combinations of ICP-AES instrumental configurations are possible. Table I outlines a few representative configurations of user interfaced and commercial systems. (The user interfaced systems are referenced.) However, this is not intended to be a complete listing of either type as such a list is not within the objective of this chapter. Nor will this discussion attempt to define what specifications and features should be included in an ICP-AES system. Specifications and design features are a function of each specific application. A fairly thorough discussion of specifications, relative to specific applications, is provided in a 24-page report by Haas, et. al.(19) The Haas report includes a set of rigorous specifications and an explanation of the rationale behind certain specifications. Additional helpful information can often be obtained from discussions with successful ICP users and by careful study of the reports appearing in the ICP Information Newsletter(20) as edited by R. M. Barnes.

The following discussion will focus upon three areas of importance to successful utilization of an ICP. The three specific areas discussed are nebulizer design, spectral rejection and computer processing capability and have significant influence upon the quality and quantity of data obtained from an ICP-AES. These three features will significantly contribute to the analytical sample rate and the accuracy of the data.

Nebulizers — Nebulizers of existing systems have been the object of extensive controversy. An ICP nebulizer can be the source of serious problems, if either the nebulizer or sample solutions are not treated properly, since most ICP nebulizers are fragile and easily clogged by excessive amounts of undissolved matter. However, the failure rate of the cross flow nebulizer is no greater than that of most FAA nebulizers. (This is

Table I-ICP-AES Instrumentation

Sources

Free running oscillators	**Detection Systems**
6KW, 7MHz	"Large quartz Spectrograph"[1]
6KW, 7MHz	3m 30 channel polychromator[15]
6KW, 7MHz	1m~30 channel polychromator[15]
6KW, 7MHz	1.5m~60 channel double polychromator[15]
2KW, 52MHz	1m Czerny-Turner monochromator[16]
2KW, 52MHz	.5m Ebert stepping motor controlled monochromator[17]
Crystal controlled oscillators	
3.0KW, 27.12MHz	1m 48 channel polychromator
3.0KW, 27.12MHz	1m scanning monochromator
2.5KW, 27.12MHz	.75m 48 channel polychromator
3.0KW, 27.12MHz	3.4m Ebert polychromator, photographic and electrical detection
2.5KW, 27.12MHz	.3m Ebert monochromator
2.5KW, 27.12MHz automatic impedance matching	.75m quartz-echelle spectrometer[18]
2.5KW, 27.12MHz automatic impedance matching	1m Czerny-Turner monochromator

indeed a feat, because the cross flow nebulizer must aspirate and nebulize two ω three milliliters of solution per minute with a one liter per minute gas flow rate, whereas an FAA nebulizer will aspirate and nebulize three to five milliliters of solution per minute with a gas flow rate of four to six liters per minute. Perhaps the reason that ICP nebulizer problems are so pronounced is that when there is a nebulizer failure in an ICP, twenty-five to thirty analytical values per minute are lost during the 40 to 120 minutes that the system is non-operational. The recalibration period for an ICP-AES system may require 30 to 50 minutes depending upon the precision level desired. In contrast, an efficient FAA facility will result in a loss of only five or six analytical values per minute and the recalibration period is only one to two minutes. Thus, a nebulizer failure for an FAA system may result in a loss of 60 to 100 determinations, whereas similar ICP failure may result in a loss of as many as 3600 determinations. Thus, it is important to minimize nebulizer failures on a multi-element ICP system. The experience of many laboratories, including the authors, serves to demonstrate that this goal can be achieved.

Five basic nebulizer designs, known to be used on ICP systems are: cross flow, concentric, flow shear, ultrasonic, and FAA. The cross flow and concentric designs are most commonly supplied with commercial systems. FAA nebulizers were supplied with early multi-element systems and there are some commercial suppliers who provide ultrasonic nebulizers.

The application of the cross flow nebulizer to ICP-AES was first described by

Kniseley, et. al.[21] We personally prefer to use, exclusively, a practical modification of the cross flow nebulizer described by Kniseley[21], which is currently available from only one source, namely, Plasma-Therm, Inc., (Cat. No. TN 5500, Jarrell-Ash recently made available a complex cross flow nebulizer.) Both the Kniseley and Plasma-Therm, Inc., cross flow nebulizers have operated for markedly extensive time periods at the Barringer Research Laboratory and have required no attention. Cross flow nebulizers are capable of handling aqueous sodium chloride in concentrations greater than 25%. The assembly and adjustment of the TN 5500 cross flow nebulizer is a relatively simple task when properly carried out and the replacement of the glass capillaries is inexpensive in the event that hydrofluoric acid is inadvertently present in a sample. The glass capillaries of cross flow nebulizers are not constricted at the tips, which minimizes clogging. However, solutions containing floating visible fibers should be filtered. Most fine particles will settle from solution if allowed to stand for one to two hours. Although particles that remain suspended may not clog the nebulizer, the particles may contribute unwanted elemental content to the analysis. The argon capillary is relatively trouble free and provides a reliable source of argon gas if metal gas lines are used instead of "plastic lines." If Teflon tape is used on gas fittings it must not be placed in contact with the flowing gas. These last two precautions will generally apply to all nebulizers employing fine glass orifices.

The concentric nebulizer, shown in Figure 2A, consists of an inner concentric capillary, containing the solution, surrounded by another capillary, and the gas flows between the two capillaries. This configuration is similar to some commercial FAA nebulizers. The concentric glass capillaries are fused together[22] to form a rigid all-glass nebulizer. The rigidity of the fused configuration will generally eliminate misalignment, but particulates in the argon, flowing in the outer channel as defined by the concentric capillary walls, can lodge at the tip, bending the long elastic central glass capillary from a co-axial position. The tip of the central glass tube is extremely fragile and must not be touched. A recessed central tube, diagrammed in modification Figure 2B, has been described[23] but has not yet thoroughly been evaluated. The constriction of the solution capillary results in clogging from any particulates present in solution. The coaxial nebulizer will also handle aqueous sodium chloride concentrations greater than 25%. In routine analysis at Barringer Research Laboratory, the sensitivity of several coaxial nebulizers was found to degrade slowly, over a one month period. Nevertheless, the operating cost, with this failure rate is only five cents per sample. It must be stated that many laboratories choose to use the coaxial nebulizer because it is easily replaced and allows the attainment of maximum sensitivity with minimal adjustment. The choice between coaxial or cross flow nebulizers may be only a matter of preference.

FAA nebulizers[24, 25] have also been used by venting 70 to 80 per cent of the argon aerosol in order to achieve a one liter per minute aerosol gas flow rate. The sturdy FAA nebulizers do not require gentle handling. Their larger capillaries are seldom clogged with particulates, but the metal components are attacked by acids, which results in a loss of sensitivity. The gas stream splitting system is susceptible to variations in the split ratio, and erratic signals may result. Excess aerosol, which generally contains acid fumes, must be properly vented away from the ICP optics and electronics. FAA nebulizers produce larger aerosols than either the cross flow or coaxial nebulizers. An impactor or aerosol sizing chamber may be essential if solute matrix interferences are to be avoided when FAA nebulizers are used in ICP systems.

Ultrasonic nebulizers, conceptually, are ideally suited for the production of aero-

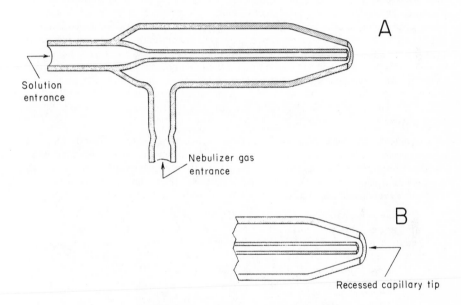

A

Solution
entrance

Nebulizer gas
entrance

B

Recessed capillary tip

Figure 2. Coaxial nebulizers, cross section

sol for ICP systems, because the function of the argon carrier gas is only transport and plasma injection. The aspiration and nebulization processes are not dependent upon argon flow, and allow the argon flow to be optimized in order to obtain plasma injection while minimizing aerosol dilution by the argon gas. The rate of aerosol formation, assuming the piezoelectric surface is covered with liquid, is dependent upon the radio frequency power driving the crystal and upon the physical properties of the liquid. The oscillation frequency, of one to three megahertz, determines aerosol size, which is usually only a few microns in range. Ultrasonic nebulizers have been used for the last eleven years[2, 26-34] but recently an increased interest has developed. Olson et. al.[33] described performance characteristics, including detection limits, of a practical ultrasonic nebulizer configuration, which was used as a model by a commercial manufacturer (Plasma-Therm, Inc., Cat. No. UNS-1). An order of magnitude improvement in detection limits, relative to pneumatic nebulization, was reported by Olson[33]. Hass et. al.[34], continuing the work of Olson et. al.[33], described the application of an ultrasonic nebulizer ICP system for the analysis of trace elements in urine. A desolvation system was used to prevent the increased liquid transport from extinguishing the plasma, and an internal standard was used to correct for signal intensity differences originating from the sample-to-sample variation of the total solids content. One advantage of previous ICP applications was the elimination of internal standardization requirements and problems associated with internal standardization. A careful evaluation of ultrasonic nebulizers may be warrented before these nebulizers are used in the analysis of actual analytical samples.

A new class of nebulizers, referred to as a "flow-shear" nebulizer, is designed to direct a gravity flow of solution across a surface containing the nebulizing gas orifice. Because only the nebulizing gas is forced through an orifice to create the aerosol, there is no solution capillary that can be clogged. With this nebulizer, the argon carrier gas only nebulizes and transports the aerosol and does not aspirate the sample solution.

The flow-shear nebulizer consists of a spherical surface with a fine slot through which the argon gas passes horizontally, and creates an aerosol stream flowing normal to the tangent at the slot. The Babington flow-shear nebulizer has been used for FAA, by Fry and Denton[36]. The Fry and Denton version requires a flow rate of nine to twelve liters per minute of nebulizing gas. It may be possible to select a proper orifice size to obtain adequate aerosol production with a nebulizing gas flow rate of one liter per minute which is more suited to most ICP systems. A peristaltic pump transfers the solution to the nebulizer.

Suddendorf and Boyer[37] have described the application of a flow-shear nebulizer for the analysis of high-salt-content samples with an ICP-AES system. The nebulizer described by Suddendorf differs from the Babington design in that the solution flows down an inclined "V"-shaped trough. The aerosol gas orifice is in the middle of the trough at the apex. The gas exits in a direction normal to the apex and nebulizes the solution against an impactor. The sample solution must be pumped to the flow shear nebulizer. Sample solutions containing high sewage sludge (1% w/v), when analyzed with the Suddendorf nebulizer, gave elemental concentrations for Al, Ca, Cu, Fe, Mn, P, and Zn which were fairly similar to the elemental concentrations obtained with a Jarrell-Ash cross flow nebulizer. Further refinements of this nebulizer may be expected because the reported detection limits are not as good as the cross flow nebulizer detection limits reported by Olson, et. al.[34]

Spectral Purity — The accuracy of any atomic spectroscopic technique is limited by many factors, one of which is the spectral purity of the radiation reaching the detector(s). This topic is an important instrumental consideration. As analytical chemists, we have been constantly challenged by more discerning individuals and agencies to verify the accuracy of ICP data. It is from these experiences that, in addition to developing confidence in the analytical system, we learned where weaknesses in the ICP system existed. The analytical results obtained during the first 18 months of using the ICP were accurate for most, but not all elements, in many sample matrices analyzed. Certain matrices (e.g. sludges and sediments) created accuracy problems for many trace elements. Any questions concerning poor accuracy, undetected by our quality assurance analysis program of reference materials, were brought to our attention by our clients. In spite of the early problems that we experienced, it is now possible to obtain accurate simultaneous multi-elemental analytical results that are within five percent of the reference value. This is a result of modifications, which will be detailed later. Before implementing these modifications it was necessary to "not report" certain trace element concentration-matrix combinations.

Larson et. al.[38] and Silvester[39] have reported that the origin of ICP accuracy problems resulted primarily from stray radiation reaching the photomultipliers. Stray radiation may be observed in spectrometers using an inductively coupled argon plasma torch because the resulting spectra exhibit excellent ratios of line intensity to background noise. The improved line-to-background ratios permit an increase in analytical sensitivity. During more sensitive analysis, stray radiation phenomena, previously masked by source noise from emission approaches with poorer detection limits, contribute erroneously to the analyte signal in a manner unexpected from studies with simple analytical solutions. For the purpose of this discussion we will define stray radiation effects as "spectral phenomena", which include radiation level changes at the photodetector that are not related to the concentration of the analyte. The three primary sources of stray radiation are spectral line overlap, surface scatter, and continuum effects. Surface scatter and spectral overlap cause erroneously high measurements, but continuum effects result in either high or low measurements.

Spectral overlap in atomic emission spectroscopy has long been recognized as a significant source of analytical error. Spectral overlap occurs when a significant fraction of a discrete emission line originating from a non-analyte element (i.e. interferent), falls within the band pass of the exit slit. In some situations spectral overlap can be avoided by (1) increased spectrometer resolution if the two emission lines are not coincident, (2) selection of an alternate analyte emission wavelength free from any other spectral overlap problems, or (3) calculating and subtracting the interference from the raw analysis data. Increasing the spectrometer resolution is ultimately limited by the actual line widths and line separations.

Previous experience in arc and spark emission spectroscopy has revealed numerous spectral overlap problems. Wavelength tables exist that tabulate spectral emission lines and relative intensities for the purpose of facilitating wavelength selection. Although the spectral interference information available from arc and spark spectroscopy is extremely useful, the information is not sufficient to avoid all ICP spectral interferences. ICP spectra differ from arc and spark emission spectra because the line intensities are not directly comparable. As of yet, there is no atlas of ICP emission line intensity data, that would facilitate line selection based upon element concentrations, intensity ratios and spectral band pass. This is indeed unfortunate because the ICP instrumentation is now capable of precise and easily duplicated intensity measurements.

The contribution of spectrometer surface scatter depends upon the design, quality, and alignment of the spectometer components. Surface scatter may originate from the grating, slits, mirrors, refractor plates, photomultiplier(s), light baffles, interference filters, dust, and the spectrometer interior surfaces. Essentially all internal surfaces may contribute to scatter.

The significance of spectrometer reflections can exemplified through the determination of the selectivity ratio required to analyze one part per million (ppm) lead in bone ash (approximately 50% or 500,000 ppm calcium) to an accuracy of ± 10 percent. To a first approximation (based upon detection limits and assumed equivalent continuum noise at the lead and calcium emission wavelengths, of 220.3 and 393.3 nm respectively) the ratio of calcium to lead photon flux is 6×10^8 for this example. The actual ratio is probably greater because we have considered only the strongest calcium line, although there are two other intense calcium emission lines. The surface scattered radiation reaching the lead photomultiplier must be less than one photon in one billion if the analytical error is to be less than ten per cent. Larsen, et. al.[38] present several illustrations of the effect of stray calcium radiation at several wavelengths. Figure 3 illustrates that stray radiation is observed at remote wavelengths, is generally without structure, and is a function of the calcium concentration.

The selectivity (or specificity ratio) is useful for defining the magnitude of an analytical interference for real situations. Photon ratios serve only to demonstrate the demands upon the spectrometer. The selectivity ratio is the concentration of interferent that causes a unit concentration error in the analyte. If the selectivity ratio of 2000 (defined as adequate by industry)[41] is used, the apparent lead concentration in the bone ash will be 250 ppm. A calcium/lead selectivity ratio of 5,000,000 is required to achieve an analytical accuracy of ± 10 per cent for one ppm lead in bone ash. (The authors are aware of a lead analysis for bone ash containing approximately 30 ppm lead that was reported by an ICP laboratory to contain approximately 550 ppm lead.) In this instance the selectivity ratio was only 1×10^3.

For simplicity, the example discussion included the effect from only one intereferent, but the interferent effects are additive. Although this may at first seem to complicate the problem, the total interferent contribution can be determined and corrected by obtaining the sum of all the interferent concentration specificity ratio response functions. This forms the basis of a stray light computer correction used at the Barringer Research Laboratory[42]. Dahlquist and Knolls[43] describe a similar computer correction approach called BLISS.

Continuum effects in the ICP have been observed at the Barringer Research Laboratory in the course of determining selectivity ratios for the computer matrix stray light correction. For certain element combinations negative selectivity ratio-concentration functions were observed. (Scattered radiation and spectral overlap can give only positive selectivity ratio-concentration functions.) The continuum effects, however, have not caused significant errors in analysis at the Barringer Research Laboratory after a computer correction for their contribution to the analysis. Horlich[44] has also reported that errors in analysis may be expected as a result of variations in molecular band emission. Horlich suggested that the band emission could be reduced by preventing air entrainment.

Various methods to achieve stray light reduction include spectrometer interior redesign; improved optical surfaces (e.g. gratings); and modifications of the dispersion

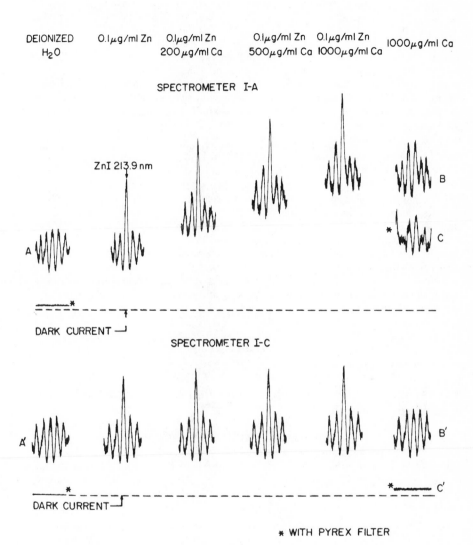

Applied Spectroscopy

Figure 3.   *Effect of scatter from calcium emission on the observed background signal in the region of the zinc, I = 213.9-nm line. Note that the borosilicate glass filter prevents any visible radiation from entering the spectrometer (38).*

process. The apparent continuum level may be monitored in the regions of interest at wavelengths where emission lines in the sample and source are absent. Scanning the spectrum by moving the entrance slit (or by moving the apparent position relative to the grating with a refractor plate) may be used to measure the background level in the vicinity of the spectral line. Placing interference filter masks over the photomultiplier windows, effectively reduce stray light originating from wavelengths that are from the analyte line by more than two times the filter band pass. Use of Notch filters[45], or band reflection filters, centered at the wavelength of the interferent radiation have been recommended to prevent the more intense radiation from entering the spectrometer.

Experience at Barringer Research Laboratory demonstrated that the most effective method to reduce stray light is to combine several reduction procedures. The lead/calcium selectivity of the instrumentation as received was 160, but this has been increased to greater than 1,000,000 with the manufacturer's modifications. The actual steps included: replacement of the calcium 393.3 nm line with the 315.9 nm line, replacement of the lead 405.8 nm line with the 220.3 nm line, installation of an interference filter mask over the lead photomultiplier, and computer correction of the residual calcium interference. The stray light reduction obtained by installation of interference filters is presented for three common concomitants in Table II. In many cases the stray light levels were less than or equivalent to the detection limit. The interference filters and the photomultiplier masks (which reduce the entrance angle to the photomultiplier to include only the receiver mirror) improved the detection limits for many elements in the array by decreasing the system band pass.

TABLE II

Interference Filter Stray Light Reduction Factors Obtained on the Barringer Research ICP.

STRAY LIGHT REDUCTION FACTOR

Interferent	Ag	As	Be	Cu	Mo	Ni	P	Pb	Ti	V	Zn
Ca	—	> 27	11	—	2	2	> 7	> 5	2	2.5	> 8
Mg	> 12	—	> 2	—	—	> 2	> 4	> 6	—	—	—
Fe	—	—	4	> 17	—	> 3	—	—	> 6	> 3	—

Chemical separation techniques can be used to reduce spectral interferences and concentrate the analyte. These techniques include solvent extraction[39] and hydride generation[39, 46, 47]. At Imperial College, the hydride generation technique is being used on a daily basis[46] for the analysis of soils, sediments, waters, herbage, and animal tissue. The solvent extraction technique is ideally suited for automated systems where the increased manipulation is carried out automatically, and a labor intensive step and sources of contamination are avoided.

Sample Dissolution — Sample introduction into most ICP systems, is by liquid nebulization. This constraint partially limits the quality of the emission analysis to be dependent on the digestion, in the case of solid samples. The fact that several elements are easily monitored simultaneously places a greater demand on the care and choice of sample preparation. Also there are both advantages and disadvantages to the use of dissolved samples in analysis. Some disadvantages are:

a) The introduction of more steps in the analytical sequence increases the chances of operator associated error and of contamination.

b) The additional dilution introduced reduces the sensitivity and detectability proportionally.

c) Sample modification may lead to loss of analyte (e.g., As, Se, Cr[48]).

d) The decomposition used may not accurately reflect the total inorganic composition, owing to incompleteness of attack.

All of these disadvantages, however, may be more apparent than real. With a combination of meticulous care, optimization, and control of the dissolution process, a good analyst should be able to overcome these hinderances.

A few advantages of dissolution are:

a) The process ensures homogeniety in the final sample form. This is especially desirable for sample types that are very difficult to homogenize prior to sub-sampling (e.g., oil filters, bird feathers).

b) The selectivity of some solid-liquid extractions that are extensively applied in exploration geochemistry can be extremely useful in environmental applications. In the study by Agemian and Chau[49] it was found that the combination 4.0 N HNO_3 — 0.7 N HCl attacks the silicate lattice of river sediments, whereas 0.5 N HCl and 0.05 N EDTA were suitable for the simultaneous extraction of Cd, Cu, Pb, Zn, Ni, Cr, Co, Mn, Fe, and Al from the organic absorbed and precipitated pahses of sediments. The latter extraction gave a measure of non-detrital or authigenic heavy metals in sediments and is thus useful to studies of environmental contamination.

The choice of a decomposition procedure for solids depends upon the type of information required. Generally the digestions for multi-element analysis would require complete destruction of the organic matrix or decomposition of the inorganic constituents. Digestions developed for FAA analysis may be used. In addition, there is a wealth of literature on various decomposition techniques for organic[50, 51] and inorganic[52] materials. Many ICP publications have reported a number of digestion procedures. The work of Dahlquist and Knoll[43] is one of the most comprehensive, which discuss the analysis of 19 trace elements in diverse biologically related samples[53]; trace elements in coal and coal fly ash[54]; fifteen trace elements in waters, soils, airborne materials, and biological tissues[55]; phosphorus in soils[56] and milk powders[57]; and fourteen elements in sewage sludge, animal feed and manure[58].

The reagents most commonly used are $HClO_4$ — HNO_3, aqua regia, HF with $HClO_4$ or H_2SO_4, $LiBO_2$, or any of the above acids in a Teflon-lined bomb. $HClO_4$ — HNO_3 (1:4) will wet-ash most organic matrixes and attack inorganic residues. Pro grammed dry ashing at 500°C for 2 hours (for organic samples), followed by aqua regia or HF digestion, may be used unless there is concern for volatile forms of trace metals. Aqua regia attacks most inorganic samples as well as those containing a large percentage of organics. HF decomposes silicates readily, owing to volatilization of Si as SiF_4.

solution preparation and relatively high levels of stray light generated by these materials. Table **III** presents ICP analytical data obtained for USGS and CANMET rock standards. The data for this table are randomly selected from the analyses of these materials as quality assurance standards placed within the analytical batches at Barringer Research Laboratory. The solutions were prepared by digesting the samples in an $HF/HNO_3/HClO_4$ mixture. More recent instrumentation and procedural refinements have occurred which refine the precision and accuracy of the data.

Applications to Environmental Samples — A number of publications are available demonstrating that ICP will give accurate results for a variety of samples (C.A. NBS orchard leaves and bovine liver), but there are still a few reports that describe applications where the ICP has been used as the tool for the evaluation of an environmental problem. Some of the applications that are discussed will be accounts from personal communications, which because of propietary reasons may be only general descriptions of the application.

Dahlquist and Knolls[50] present the most comprehensive discussion of ICP analysis for metals in biological samples. Greenfield[7, 15] describes environmental problems in his ICP application studies. The primary environmental applications reported are particulates from stacks and ambient air. Ronan and Kunselman[72] compared ICP results with FAA values for 23 elements in natural waters with emphasis on cost and sample volume. Winge et. al.[73] discussed the instrumental requirements for the analyses of natural waters, focusing upon stray light effects. Silvester[39] reported on the limitations and advantages of ICP analysis for 25 elements in natural waters using a first generation commercial instrument coupled with techniques of direct aspiration, solvent extraction, hydride formation, and electrothermal volatilization. Detection limits, linearity, and interferences are discussed.

The analysis of soils and plant material are common examples used to demonstrate ICP applications. Dahlquist and Knoll[43] compared the preparation and ICP analyses of botanicals (16 elements) and soils (11 elements); with few exceptions the ICP values for the CII botanicals were in excellent agreement with the assigned values, and the soil analyses were in excellent agreement with FAA analyses of soil digests. Jones[74] reported the analysis of 17 elements in plant material and soils but confirmation of the two analyses was not given. Alder, et. al.[75] describe the unique analysis of ammonia-nitrogen in soils by gas evolution into an ICP; no interferences were observed from the concomitants evaluated and acceptable recoveries were obtained. Irons et. al.[76] compared the ICP analyses of 13 elements in NBS orchard leaves and bovine liver to the data obtained by FAA and energy dispersive x-ray.

Fortescue et. al.[77] described the environmental applications of ICP and FAA analysis to sewage sludges generated by seven Niagra Penninsula communities. The results obtained by the two techniques were compared, and contributions from individual industrial sources were described. A concern was expressed cautioning the application of certain sludges to agricultural areas as fertilizer. Taylor[78] has also compared the ICP analysis of sewage influent and effluent using neutron activation and atomic absorption analysis techniques for 26 elements. With a few exceptions (antimony, colbalt, molybdenum, and vanadium) excellent agreement was obtained.

Capar et. al.[59] described an environmental application in which ICP was one of four instrumental approaches employed to evaluate a potential environmental problem. The authors also explain that sewage sludge is being considered for feed wastelage. The objective of the Capar et. al.[59] study is the determination of the health

TABLE III

COMPARATIVE DATA FOR U.S.G.S. AND C.R.M.P. ROCK STANDARDS (TRACE ELEMENTS, PARTS PER MILLION)*

	Ba		Zn		Ni	
	BRL ICP	Certified	BRL ICP	Certified	BRL ICP	Certified
AGV-1	1290 ± 41	1410	110 ± 2	112	15 ± 6	17.8
BCR-1	695 ± 80	790	137 ± 7	132	16 ± 2	15.0
G-2	1940 ± 12	1950	89.6 ± 5	74.9	7 ± 2	6.4
GSP-1	1390 ± 150	1360	144 ± 4	143	7 ± 2	10.7
SY-2	508 ± 80	450	258 ± 10	250	11 ± 3	10

	Cu		Sr		V		Be	
	BRL ICP	Certified	BRL ICP	Certified	BRL ICP	Certified	BRL ICP	Certified
AGV-1	61.7 ± 3.7	63.7	645 ± 40	657	116 ± 7	121	2.2 ± 0.1	1.8
BCR-1	22.1 ± 4.0	22.4	315 ± 2	345	362 ± 9	384	2.9 ± 0.4	2.6
G-2	10.7 ± 0.4	10.7	458 ± 8	463	38 ± 1	37.0	1.9 ± 0.1	2.4
GSP-1	36.6 ± 0.9	35.2	243 ± 14	247	56 ± 1	52.0	1.3 ± 0.1	0.8
SY-2	5.4 ± 1.4	5	242 ± 14	250	50 ± 1	50	22.0 ± 3.5	20

* Mean of four replicate analyses

automated ICP operation. The computer can be programmed to carry out the following tasks with an automated program:

a) run reagent blank
b) subtract reagent blank automatically
c) convert raw data to ppm solution
d) calculate spectral corrections
e) store both spectral corrected and uncorrected data
f) recalibrate as required
g) encode the quality control samples for separate future retrieval
h) run dilutions of samples, as necessary
i) maintain full control of peripherals: auto sampler and ICP system

Interferences — The singular advantage of the ICP emission source above others is the absence of chemical interferences. This is generally attributed to the extremely high temperatures obtaining in the center core of the torodal plasma. There have been a few reports of ionization interferences[68, 30]. We have confirmed that these effects are real but they are not serious. Ionization effects can be minimized by optimization of r.f. power, gas flow, observation height, and vertical spatial observation window imaged onto the slit, as suggested by the work of Edmonds[69] and by experiences related by other workers[18]. Edmonds[69] presents data (cf. Figure 4.) that facilitate a fundamental understanding of the effect of the various physical parameters associated with the ICP. The position of the emission maximum for the vertical emission profile will be shifted dramatically by any variation of these physical parameters. A small vertical spatial observation window will result in a dramatic spectrometer signal change as the maximum shifts relative to the window. A large vertical spatial observation window (5 mm) will result in essentially no response change, because the window effectively integrates the entire area under the maximum. At the appropriate observation height with a 5 mm window, profile maxima will remain within the window for normal variations in the physical parameters. Horlich[70] has commented upon the fact that the emission profile maximum also shifts as the salt content of the solution changes. Again in this instance a small vertical window will result in a significant spectrometer response change. This work may explain why divergent opinions about interferences prevailed during the ICP development years.

Greenfield et. al.[71] observed a reduction of signal intensity that correlates with sample intake effects from the modified solution viscosity and/or surface tension of mineral acids. This, coupled with peristaltic pumping of solutions into the nebulizer, considerably reduces physical interferences. Increased salt concentration also has an effect on solution physical properties. In the experience of these authors, the high levels of salt in the matrix also increases the noise from the nebulizer system. This degradation of nebulizer performance, which is not necessarily accompanied by a proportional reduction in sensitivity, is the cause of the observed deterioration of detection limits in real samples as opposed to ideal solutions.

Accuracy, Precision, and Detection Limits — Analytical cost, accuracy, precision, and detection limits are the four main evaluation criteria for selecting an analytical method. Detection limit information will not be given here, as it is easily obtained from the literature or from instrument manufacturers. The analysis of NBS orchard leaves and bovine liver is often used to demonstrate the accuracy and precision of ICP analysis. We feel that the analysis of United States Geological Survey (USGS) and Canadian Centre for Mineral Energy Technology (CANMET) standard rocks, is a more rigorous test of ICP analytical accuracy because of the resistance to sample

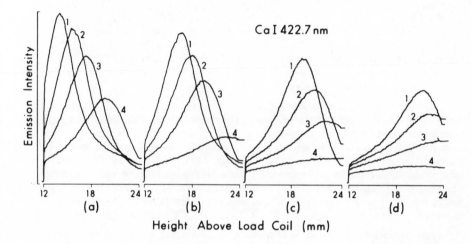

CaI 422.7 nm

Height Above Load Coil (mm)

Figure 4. Spatial profiles of calcium, I = 422.7-nm emission as a function of central axial (nebulizer) flow rate, (a) 0.9 L/min; (b) 1.0 L/min; (c) 1.1 L/min; (d) 1.2 L/min, and plasma power (curve 1, 2 kW; curve 2, 1.75 kW; curve 3, 1.5 kW; curve 4, 1.25 kW) (69).

Care must be taken to ensure complete removal of HF (or effective removal by H_3BO_3 complexation), as any free acid degrades the glassware of the nebulizer system. $LiBO_2$ is a flux that decomposes most silicates, keeping SiO_2 in solution for analysis.

The slowest step, which determines the speed of the entire analysis, is the sample decomposition step. Digestion of organic materials has generally taken the longest time of all sample types. Although it is true that most of these procedures could be made very efficient in batch operations, efforts to reduce preparation time by modification of procedures could prove invaluable to a routine laboratory[59]. Some examples of this as applied to the analysis of bovine liver[60] and orange juice[61] can be found.

Water samples deserve some special attention despite of the apparent simplicity of their analysis. One reason is that, depending upon the source, water samples will have varying degrees of particulates. Most pollution surveys require filtration through a 0.45 μ Millipore filter. In the event that filtration is omitted deliberately or unknowin gly, such particulates are entrained into the plasma, dissociated, and excited in the intense heat of the source. The efficiency of entrainment and nebulization depends on the specific nebulizer used, as well as the particle-size distribution. These in turn effect the degree of dissociation in the plasma. Thus, it is important to ensure that water samples are properly prepared.

References to sample contaimination in the literature are numerous. Abercrombie, et. al.[54] have noted that the information obtainable from ICP emission analysis warrants additional care because of the increased number of elements monitored. For example, if one were looking for boron and sodium, the use of glassware should be avoided. Certain digestion vessels that are presumed to be cleaned may not be acceptable. Polyethylene bottles (acid-soaked), for example, have been shown to continue releasing zinc from the matrix for as long as seven days[62]. A number of trace elements may be leached from commercial disposable test tubes. Some of the more important contaminants include Ba, Sr, Zr, Na, K, Al, Ca and Mg.

Alternate Sample Introduction — Obviously, elimination of the sample dissolution stage would greatly reduce analytical time, as it is the slowest step in the analytical scheme. Pulsed-laser vaporization using a CO_2—TEA laser seems promising[63, 64]. Another possibility is the introduction of a suitable prepared slurry of the sample into the nebulizer[65]. Thermal vaporization studies using heated substrates such as tantalum[66], carbon filaments[67], or carbon rods[39] have been reported. Silvester[39] de fined the problems of vapor transport, carrier gas expansion, and solid phase chemistry associated with electrothermal sample introduction to an ICP.

Data Acquistion and Processing — The minimum requirement of a data-handling system for ICP-AES is to receive the multi-parameter signals and to transform the signals into hard copy. A suitable computer system should have enough memory to store calibration coefficients, perform calculations and store the partly processed data (e.g., ppm abundance of 10 to 30 elements in the sample solution), into a storage medium (i.e., magnetic tape, paper tape, floppy disk, hard disk, or punched cards). The data can then be conveniently retrieved for further processing such as generating analytical report forms of sample-element concentrations, data quality assurance review, matrix interference corrections, generation of graphical representations, and statistical data interpretation. A small computer with 8K word read-write memory will just handle these tasks in a scheduled manner, i.e., data acquisition during the day and data processing at night for a 20-element array. Increasing the memory to 16K will increase the data processing efficiency by enabling real-time matrix correction and

related aspects of using "wastelage" (A mixture of feed and manure) for feeding animals. The trace element levels in wastelage were significantly higher than the levels expected in normal feeds. These elevated levels might lead to increased uptake if present in the animal diet, but more detailed studies including animal tissue analyses are required to draw further conclusions.

Results[79] from a Round Robin analyses for the lead content of air particulate filters were given along with the average values obtained from 65 laboratories participating in the study. The procedures used by the other laboratories included x-ray, FAA, flameless atomic absorption, emission spectrometric, and dithizone-spectrophotometric. The overall analytical agreement was very good.

McQuaker et. al.[56, 80, 81] in three sequential papers described the ICP digestion procedure development calibration, and application for 28 elements from a moderate soil and air filter sampling program. A potential source of airborne particulates containing toxic heavy metals was located at the study site. The control and study sampling sites were separated by 50km with geographic features presenting a situation that minimized mixing of the airsheds. ICP detection limits were more than adequate to monitor compliance for British Columbia[82] air quality levels. Heavy metal enrichments of airborne particulates relative to soil materials were observed at the study site, which were consistent with the expected heavy metal content of the particulate emission source.

Abercrombie et. al.[63] have described an airborne particulate collection and ICP analysis system for air quality monitoring or plume dispersal study programs. A mobile particulate collector mounted in a vehicle or aircraft was used to collect duplicate samples of six size-fractioned samples per minute. The airborne system was able to transect the plume at several altitudes and distances from the source within a few hours, minimizing the effect of changing atmospheric conditions upon the study. The airborne particulates were impinged upon the adhesive surface of a mylar tape which yielded discrete 3mm sample spots. The sampler was sequentially advanced during collection. The analysis was carried out in the laboratory using a sequential tape drive, a CO_2 laser for sample volatilization, and 25 element direct reading ICP interfaced to a desk top programable calculator. In the analysis mode, the sample spots on the tape were sequentially transported to the focus of the pulsed laser. The laser vaporized sample was entrained into an argon stream and swept into the ICP. Particle size dependant fractions were collected which made it possible to determine the sedimintation of particulates in the plume and the effectiveness of stack particulate control devices.

Thompson et. al.[46, 83, 84] have reported the ICP analysis of hydride forming elements in soils, sediments, rocks, waters, herbage, and animal tissue. In preliminary studies, the chemical conditions for forming hydrides of arsenic, antimony, bismuth, selenium and tellurium are presented. The effect of interferences and procedures for the reduction of interferences are discussed. The techniques described in this paper can be applied to environmental analysis. The primary difference between environmental and geochemical analysis occurs only in the interpretation and application of the data. Single element analytical techniques generally preclude gathering apparantly extraneous data. For multi-element ICP systems, as with many other multielement techniques, no extra effort is required to obtain the information for extraneous elements in the analysis program. Schuetzle[85] has emphasized the value of multi-element analysis for environmental studies. "We use this technique (ICP) for most analyses even though quantitative data on only one or two elements are needed. Very often, the extra data obtained during these analyses proves to be invaluable. For example, the value of obtaining this type of analysis was demonstrated in a recent air pollution study. For the past three years we have been collecting air particulate samples in the Allegheny Tunnel in Pennsylvania. Trace element analysis has been performed by FAA on six or seven elements which are used to determine the soil, car and truck contribution to the total particulate mass. However, we had been unable to accurately determine the soil contribution until we noticed that a number of other trace elements which had been generated as a result of ICP analysis gave excellent correlations with the trace element composition of the surface soils."

Summary — The ICP has been successfully used to provide bulk elemental analyses of a large variety of environmental samples. (In chapter 9 of this volume Linton discusses some advantages of surface analysis relative to bulk chemical analysis). Multi-element ICP analysis can be used for a large scale sample volume program analyzing 200 to 300 samples per day. ICP has the combined analytical attributes of high detection limits, high accuracy and precision and low cost. The ICP is becoming an altractive alternative to other trace-element analytical systems such as x-ray fluorescence, FAA, neutron activation, and arc and spark emission spectroscopy for environmental as well as other analysis.

Acknowledgements — The author greatfully acknowledge the contributions of Alice Abercrombie, Arlene Spear, Dr. Ralph King and Dr. Edward Bingler to the preparation of this manuscript.

REFERENCES

(1) Greenfield, S., Jones, I., Berry, C. T., *Analyst (London)* (1964), **89,** 713.

(2) Fassel, V. A., Wendt, R. H., *Anal. Chem.* (1965), **37,** 920.

(3) Fassel, V. A. "Electrical Flame Spectroscopy" in Proceedings XVI Colloquium Spectroscopicion Internationale, Adam Hilger, London, 1971, pp. 63-93.

(4) Greenfield, S., *Proc. Anal. Div. Chem. Soc.,* (1976), **13,** 279.

(5) Greenfield, S., McGeachin, H.McD., Smith, P. B., *Talanta,* (1976), **23,** 1.

(6) Fassel, V. A., ASTM Spec. Tech. Publ., (Flameless At. Absorpt. Anal; Update, Symp.), (1977) 22, 618.

(7) Greenfield, S., *The Spex Speaker,* Metuchen, N. J., (1977), Vol. XXII, 1.

(8) Fassel, V. A., and Kniseley, R. N., *Anal. Chem.,* (1974), **46,** 1110 A.

(9) *Ibid.,* 1155 A.

(11) Mermet, J. M., *Spectrochim Acta,* (1975), **30B,** 383.

(12) Veillon, C. and Margoshes, M., *Spectrochim Acta,* (1968), **23B,** 503.

(13) Wendt, R. H. and Fassel, V. A., *Anal. Chem.,* (1966), **38,** 337.

(14) Konnblum G. R. and de Galan, L., *Spectrochim, Acta,* (1977), **32B,** 455.

(15) Greenfield, S. Jones, I. LL., McGeachin, H. McD., Smith, P. B., *Anal. Chem. Acta,* (1975), **74,** 225.

(16) Boumans, P. W. J. M. and deBoer, F. J., *Spectrochim, Acta,* (1972), **27B,** 391.

(17) Boumans, P. W. J. M., van Goal, G. H., Jansen, J. A. J., *Analyst (London)*, (1976), **101**, 585.

(18) Dube, G., and Boulos, M. I., *Can. J. Spectrosc.*, (1977), **22**, 68.

(19) Haas, W. J. Jr., Winge, R. K., Kniseley, R. N., Fassel, V. A., NTIS-U.S. Dept. of Commerce PB-154353, (1978).

(20) Barnes, R. M., ed., *ICP Information Newsletter*, University of Massachusetts, Amherst, Mass.

(21) Kniseley, R. N., Amenson, H., Butler, C. C., Fassel, V. A., *Appl. Spectrosc.* (1974), **28**, 285.

(22) Scott, R. A., *ICP Information Newsletter*, (1978), **3**, 425.

(23) Bogdain B., *ICP Information Newsletter*, (1978), **3**, 491.

(24) Boumans, P. W. J. M., deBoer, F. J., Dahmen, F. J., Hoelzel, H., Meier, A., *Spectrochim. Acta*, (1975), **30B**, 499.

(25) Brech, F., Crawford R., Jarrell-Ash Div. Report, "The Jarrell-Ash Plasma Atom Comp" 590 Lincoln St., Waltham, Mass. 02154 (1975).

(26) Hoare, H. C., Mostyn, R. A., *Anal. Chem.*, (1967), **39**, 1153.

(27) Fassel, V. A., Dickinson, G. W., *Anel. Chem.*, (1968), **40**, 247.

(28) Boumans, R. W. J. M. and deBoer, F. J., *Spectrochim. Acta*, (1972), **27B**, 391.

(29) Greenfield, S., Jones, I. LL., McGeachin, H. McD., Smith, P. B., *Anal. Chem. Acta*, (1975), **84**, 225.

(30) Boumans, R. W. J. M. and deBoer, F. J., *Spectrochim. Acta*, (1975), **30B**, 309.

(31) Boumans, R. W. J. M. and deBoer, F. J., *Spectrochim. Acta*, (1976), **31B**, 355.

(32) Boumans, R. W. J. M. and deBoer, F. J., *Spectrochim. Acta*, (1977), **32B**, 365.

(33) Olson, K. W., Haas, Jr., W. J., Fassel, V. A., *Anal. Chem.*, (1977), **49**, 632.

(34) Haas, Jr., W. J., Fassel, V. A., Grabau, IV, F. G., Kniseley, R. N., Sutherland, W. L., paper presented at Ultratrace Metal Analysis in Biological Science and Environment Symposium, 174th ACS National Meeting, Chicago, Ill., Aug. 28-Sept. 2, (1977).

(35) Babington, R. S., U.S. Patents 3, 421, 692; 3, 421, 699; 3, 425, 058 and 3, 425, 059.

(36) Fry, R. C. and Denton, M. B., *Anal. Chem.*, (1977), **49**, 1413.

(37) Suddendorf, R. F., Boyer, K. W., Paper 395 presented at the 29th Pittsburgh Conference on Analytical Chemistry and Applied Spectroscopy, February-March 1978.

(38) Lason, G. F., Fassel, V. A., Winge, R. K., Kniseley, R. N., *Appl. Spectros.*, (1976), **30**, 384.

(39) Silvester, M. D., Department of Supply and Services Report SS04. 23233-5-1674, Hull, Quebec, March (1976).

(40) Danielson, A., Lindblom, P., *Appl. Spectrosc.*, (1976), **30**, 151.

(41) Breck, F., *ICP Newsletter*, (1976), **1**, 245.

(42) Abercrombie, F. N., Silvester, M. D., Paper 168 presented in part at 27th Pittsburgh Conference on Analytical Chemistry and Applied Spectroscopy, March (1976).

(43) Dahlquist, R. L., Knoll, J. W., *Appl. Spectros.*, (1978), **32**, 1.

(44) Horlich, G., *Ind. Res. & Dev.*, (1978), **20**, 70.

(45) Fassel, V. A., Katzenberger, J. M., Winge, R. K., Submitted (1978).

(46) Thompson, M., Pahlavanpour, B., Walton, S. J., Kirkbright, G. F., *Analyst*, (1978), **103**, 705.

(47) Fry, R. C., Denton, M. B., Windsor, D. L., Northway, J. S., "Arsenic Hydride Preconcentration for ICP-AES", Office of Naval Research, Technology, Report No. 13, (1977).

(48) Parr, R. M., *J. Radional. Chem.*, (1977), **39**, 421.

(49) Agemian, H., Chau, A. S. Y., *Arch. Environm. Contain. Toxicol.*, (1977), **6**, 69.

(50) Hanson, N. W., *Official Standardised and Recommended Methods of Analysis*, SAC, London, 1973.

(51) Smith, G. Fredrick, *The Wet-Chemical Oxidation of Organic Compositions, Employing Perchloric Acid*, G. F. Smith Chemical Co., Columbus, Ohio, 1965.

(52) Sulcek, Z., Povondra, P., Delezal, J., *Decomposition Procedures in Inorganic Analysis*, Critical Reviews in Anal. Chem., Vol. 6, Iss3, 1977, CRC.

(53) Abercrombie, F. N., Silvester, M. D., Cruz, R. B., Ch. 2, Advances in Chemistry Series, Symposium on Ultratrace Metal Analysis in Biological Sciences and Environment, Ed. T. Risby, accepted.

(54) Ward, A. F., Marciello, L., Paper 391 presented by the 29th Pittsburgh Conference on Analytical Chemistry and Applied Spectroscopy, Cleveland (1978).

(55) McQuaker, N. R., Brown, D. F., Kluckner, P. P., Submitted 1978.

(56) Gunn, A. M., Kirkbright, G. F., Opheim, L. N., *Anal. Chem.*, (1977), **40**, 1492.

(58) Capar, S. G., Tanner, J. T., Friedman, M. H., Boyer, K. W., *Environ. Sci. Technol.*, (1976), **10**, 683.

(59) Zasoski, R. J., Buran, R. G., *Communic. Soil Sci. Plant Anal.*, (1977), **8**, 425.

(60) Hinners, T. A., *Z. Anal. Chem.*, (1975), **277**, 425.

(61) McHard, J. A., Winefordium, J. D., Attaway, J. A., *J. Agric. Food Chem.*, (1976), **24**, 41.

(62) Capocianco, V., Agemian, H., Canada Center for Inland Waters, private communication.

(63) Abercrombie, F. N., Silvester, M. D., Barringer, A. R., Invited paper presented at Eastern Analytical Symposium, New York, (1977). Accepted for the proceedings, Barnes, R. M., ed., Franklin Publishing House.

(64) Abercrombie, F. N., Barringer, A. R., Bradshaw, P. M. D., Cruz, R. B., Murray, A. D., Invited paper presented at Geoanalysis 78, A Symposium on the Analysis of Geological Materials Sponsored by the GSC and CANMET, Ottawa, Canada, May (1978) accepted for proceedings.

(65) Fuller, C. W., *Analyst*, (1976), **101**, 961.

(66) Nixon, D. E., Fassel, V. A., Kniseley, R. N., *Anal. Chem.*, (1974), **46**, 210.

(67) Dahlquist, R. L., Knoll, J. W., Hoyt, R. E., paper presented at 21st Canadian Spectroscopy Conference, Ottawa, (1974) available as a report from Applied Research Laboratories, P.O. Box 129, Sunland, Ca. 91040.

(68) Larson, G. F., Fassel, V. A., Scott, R. N., Kinseley, R. N., *Anal. Chem.*, (1975), **47**, 233.

(69) Edmonds, T. E., Horlick, G., *Appl. Spetrosc*, (1977), **31**, 536.

(70) Horlick, G., University of Alberta Edmonton Alberta, Canada, personnal communication (1977).

(71) Greenfield, S., McGeachin, H., Smith, P. B., *Anal. Chem. Acta*, (1976), **84**, 67.

(72) Ronan, R. J., Kunselman, G., *NBS Spec. Publ. (U.S.)*, (1977), **464**, 107.

(73) Winge, R. K., Fassel, V. A., Kniseley, R. N., Dehalb, E., Hass, W. J., Jr., *Spectrochim. Acta, Part B*, (1977), **32**, 327.

(74) Jones, J. B., Jr., *Communic. Soil Sci. Plant Anal.*, (1977), **8**, 349.

(75) Alder, J. F., Gunn, A. M., Kirkbright, C. F., *Anal. Chem. Acta*, (1977), **92**, 43.

(76) Irons, R. D., Schenk, E. A., Giauque, R. D., *Clin. Chem.*, (1976), **22**, 2018.

(77) Fortescue, J. A. C., Silvester, M. D., Abercrombie, F. N., "The Chemical Composition of Sludge from Six Sewage Plants in the Niagra Penninsula, Ontario, Canada as Determined by Atomic Absorption and Emission Spectroscopy Using an Inductively Coupled Plasma Source", Proceedings of the 9th Annual Conference on Trace Substances in Environmental Health. Columbia, Missouri, June, (1975).

(78) Taylor, C. E., Environmental Protection Agency Report 600/2-77-113, June, (1977). Rogers, private communications.

(80) McQuaker, N. R., Kluckner, P. D., Chang, G. N., Submitted (1978).

(81) McQuaker, N. R., Brown, D. F., Kluckner, P. D., Submitted (1978).

(82) Pollution Control Board, "Pollution Control Objectives for the Mine, Mine-Milling and Smelting Industries of British Columbia", Province of British Columbia, 1974.

(83) Thompson, M., Pahlavanpour, B., Walton, S. J., Kirkbright, G. F., *Analyst*, (1978), **103**, 568.

(84) Thompson, M., Pahlavanpour, B., Walton, S. J., Kirkbright, G. F., In Press.

(85) Schuetzle, D., private communication, Scientific Research Lab, Ford Motor Company, Dearborn, Mi. (1978).

RECEIVED November 17, 1978.

Surface Microanalytical Techniques for the Chemical Characterization of Atmospheric Particulates

R. W. LINTON

Venable and Kenan Laboratories, University of North Carolina, Chapel Hill, NC 27514

Introduction

A prominent example of the recent, large scale mobilization of chemicals in the environment is the introduction of pollutant particles into the atmosphere. Such pollution is the direct consequence of the acquisition and processing of raw materials required to sustain advanced technological societies (1). Many elements now show substantial enrichments over natural background levels in the atmosphere (2-6). Further, nearly all particles produced by anthropogenic sources contain higher specific concentrations ($\mu g/g$) of some trace elements than natural windborne particles including crustal dust, volcanic ash, and sea salt aerosol (1, 7-9). It is becoming more evident that evolution often does not provide effective homeostatic mechanisms that permit organisms to tolerate sudden exposures to chemicals that have been previously unavailable because of their low abundance or geochemical stability (10). In the case of atmospheric particles, a major scientific concern, therefore, is to assess the extent to which trace element enrichments may result in deleterious effects on health.

The specific long-term environmental effects of increased trace element loading of the atmosphere continue to be difficult to assess. Specific areas of uncertainty requiring further investigation include the following: 1) the mechanisms of particle formation and dispersion in the environment, 2) the chemical transformations and reactivity of the particles in various environmental compartments, 3) the physico-chemical characteristics of individual particles, and 4) the specific interactions of the particles with living organisms (11).

Conventional studies generally involve the collection of an assemblage of airborne particles followed by determinations of the *average or bulk* concentrations of pollutant species present (12). However, the results often lack the analytical specificity required to identify particle sources, to determine particle speciation and reactivity, or to assess particle toxicity.

An obvious limitation to the use of bulk analysis studies is the direct result of sample heterogeneity. Not only do aerosol samples show wide variability in the physico-chemical characteristics of different particles, but even a single airborne particle may be highly heterogeneous. With regard to the latter, the surface chemical composition of a particle may bear little resemblance to that of its interior (11-14).

0-8412-0480-2/79/47-094-137$05.75/0

The surface composition of individual airborne particles is of particular importance for the following reasons:

1. A number of potentially toxic trace metal and organic species are highly enriched at the surfaces of many types of environmental particles (11-14).

2. It is the surface of the particle that is directly accessible to extraction by aqueous leaching in the environment or by body fluids following inhalation or ingestion (11, 12, 14).

3. The particle surface may function as a catalytic site for heterogeneous reactions involving the generation or removal of gaseous pollutants (11, 15-17).

The physical characteristics of individual particles also are of environmental significance. For example, the smaller particles (diameters on the order of 1 micrometer of less) generally are most important in that they have very long atmospheric residence times (18), are least effectively controlled by pollution control devices (19), are preferentially deposited in the pulmonary regions of the lung (20, 21), and may be most enriched in toxic species on a specific concentration (μg/g) basis (22-24).

The above considerations clearly point to the need for *surface microanalytical* techniques that allow for the *direct* determination of the physical and chemical composition of individual particles (11). The purpose of this paper, therefore, is to review modern analytical advances in this area.

Analytical Instrumentation

Recent technological innovations have permitted the rapid emergence of an array of spectroscopies used for the determination of the composition and microstructure of the outermost atomic layers of a solid (25, 26). There are three major spectroscopic techniques that combine both microscopic and surface analysis capabilities, and that are beginning to find useful applications in the characterization of environmental microparticles. These are electron excited X-ray emission spectroscopy (electron microprobe, or scanning electron microscope equipped with an X-ray detector) (27, 28), Auger electron spectrometry (scanning Auger microprobe) (29) and secondary ion mass spectrometry (ion microprobe) (30). All three techniques have comparable lateral resolutions, i.e. features as small as about 1 micrometer in diameter may be characterized. However, the spectroscopies differ considerably with regard to surface sensitivity and specificity.

The electron microprobe (EMP) and scanning Auger microprobe (SAM) respectively monitor the emitted X-rays and Auger electrons that result from electron bombardment-induced core ionization of the sample atom (Figure 1). The excess energy lost by an outer shell electron that fills the core vacancy can be emitted as either an X-ray, or imparted to another outer shell electron (the Auger electron) causing its ejection from the atom (Figure 1). The energy of the emitted X-rays and Auger electrons are characteristic of the emitting elements. Since Auger electron emission predominates over the X-ray emission process for low atomic number elements, the Auger technique usually is more sensitive than the EMP for light elements (Li → Na). For heavier elements, the sensitivities of both techniques are roughly comparable and are on the order of 0.1% atomic percent within the analytical volume.

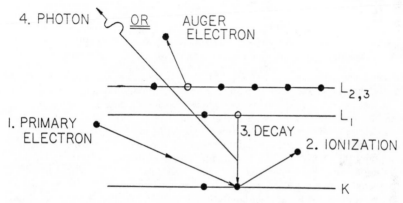

Figure 1.   *Schematic demonstrating competitive auger electron and x-ray photon emission processes*

Core vacancies resulting from electron bombardment are achieved at depths ranging up to micrometers into the sample. The X-rays subsequently produced are able to traverse this depth range and thus the depth resolution of the EMP is poor. Micrometer-sized analysis depths are often much greater than the immediate surface layer normally of interest in the case of environmental particles (11-14). On the other hand, the Auger electrons produced can travel only short distances in the solid without energy loss. Consequently, only the Auger electrons originating within a few atomic layers of the surface (generally <20 Å) will be ejected from the solid without undergoing energy loss and will be detected. The Auger technique, therefore, has excellent surface specificity.

Because many environmental particles are poor electrical conductors, charging produced by the incident electron beam is a major analytical concern in EMP and SAM (11, 14). Because EMP analysis is not very surface specific, the sample surface is normally coated with a thin film of a low Z conducting material such as C. The surface specificity of the Auger technique generally precludes the use of surface coatings, and electrical charging of the particles may be exceedingly difficult to overcome experimentally (11, 14).

Examples of conventional instrumentation used for electron-excited X-ray emission spectroscopy and Auger electron spectrometry are shown in Figures 2 and 3 respectively. Details concerning the instrumentation may be found elsewhere (25-29).

In secondary ion mass spectrometry (SIMS), a beam of energetic primary ions is used to eject surface species from a solid sample. The primary ion-surface interaction may be conceptualized as a series of hard-sphere collisions (Figure 4). The primary ion loses energy during successive atomic collisions and comes to rest (i.e., is implanted) at depths on the order of a few hundred angstroms, As illustrated (Figure 4), a series of angstroms glancing collisions in the solid is required to eject or sputter surface species. However, only the atoms within a few atomic layers of the surface may be ejected following the impact of a primary ion. Although only a small fraction (often less than 1%) of the sputtered species are ionized, they can be analyzed with a mass spectrometer to provide a sensitive surface analysis. Elemental detection limits range from about 10^{-2} to 10^{-6} atomic percent depending primarily upon the element and primary ion beam conditions employed. Typically, it is possible to observe as little as 1 μg/g in the analytical volume thereby enabling studies of surface elements present at trace levels (11, 14, 25, 26, 30).

The ion microprobe mass spectrometer represents a special configuration of SIMS in which the primary beam can be focused to diameters as small as 1 μm (Figure 5). Spectral interferences from molecular and multiply-charged ions make the high resolving power of a double focusing mass spectrometer (e.g. Figure 5) highly desirable, especially for chemically complex matrices such as those characteristic of environmental samples. Details about the sputtering process and ion microprobe instrumentation are available elsewhere (25, 30, 31).

Although not capable of the micrometer-sized lateral resolutions available with the aforementioned techniques, the surface spectroscopy, electron spectroscopy for chemical analysis (ESCA), also deserves mention. The ESCA experiment involves the use of X-rays rather than electrons to eject core electrons (photoelectrons), and it has comparable surface specificity and sensitivity to that of Auger electron spectroscopy (AES) (25, 26, 29). The principal advantage of ESCA relative to AES is that small

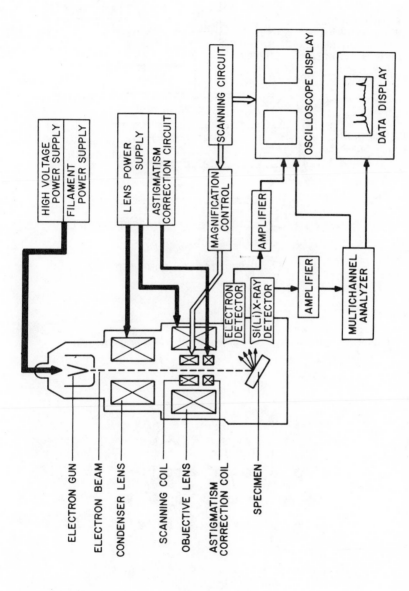

Figure 2.   Schematic of a scanning electron microscope equipped for energy dispersive X-ray analysis

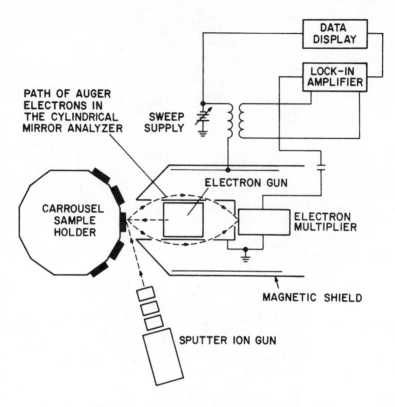

Figure 3.   Schematic of a conventional auger electron spectrometer

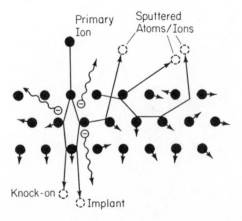

Figure 4.   The sputtering process—interaction of primary ions with a sample surface

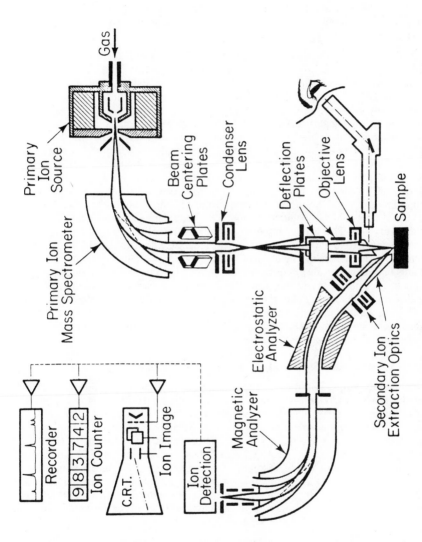

Figure 5. Schematic of a conventional ion microprobe mass spectrometer

changes in the observed binding energies of the photoelectrons can be interpreted on the basis of simple valence bond concepts and thus can often provide information about surface chemical *speciation* (29). For example, a shift to a higher binding energy often reflects a shift to a higher oxidation state. Unfortunately, the utility of ESCA for individual particle analysis is very limited owing to the difficulty of focusing the exciting X-ray beam submillimeter diameters. Recent advances, however, suggest that beam to lateral resolutions on the order of 10 μm are feasible using novel X-ray source designs (32, 33).

Before specific environmental applications of the surface techniques are discussed, it is important to note that there are some general problems, not normally encountered in bulk measurements, which are associated with surface microscopic analysis (11, 12, 14). First, quantitation of elemental concentrations is much more difficult largely because of the uncertainty in defining a precise analytical volume, and the difficulty in preparing standards suitable for surface specific measurements. A second concern involves the inherent trade-off between analytical specificity and statistics. In achieving the specificity associated with single particle analysis, one does sacrifice the ability to obtain a representative measurement of the surface composition of the bulk sample, unless a large number of particles are studied sequentially or a group of particles is analyzed at once. A third problem is that *absolute* elemental detection limits (in g) of the surface techniques must be much better than those of the conventional bulk analysis techiques to obtain similar *relative* or concentration detection limits (in μg/g). This is the consequence of the extremely small volumes sampled with a surface microanalytical technique. To summarize, although the absolute detection limits of a surface technique may be extraordinarily good (10^{-15}g), relative detection limits only may be 1,000 μg/g within the analytical volume. Finally, one must be aware of possible artifacts resulting from excitation processes that involve high energy charged-particle bombardment of the sample surface. For example, such artifacts may alter sample composition as the consequence of surface chemical reactions, the selective volatilization of elements, or the migration of ions in the bulk solid due to electrical charging at the solid surface (34, 35).

Surface Microanalytical Studies of Environmental Particles

Elemental Mapping Studies. The electron microscope or microprobe (EMP), the scanning Auger microprobe (SAM), and the ion microprobe (IMP) all employ focused beams of charged particles. The exciting beams, therefore, can be electrically or magnetically deflected and rastered over a field of particles (analogous to that of an electron beam raster in a television screen). The raster pattern of the ion or electron beam is synchronized with the electron beam raster in an oscilloscope (CRT) display. The spectrometer is tuned to monitor an X-ray, Auger electron, or secondary ion emission characteristic of the element of interest. The intensity of the emission then is used to control the intensity of the CRT display. A bright area on the CRT thus corresponds to a high concentration of the element at that location in the analytical field.

It is obvious that this "elemental mapping" capability can be highly advantageous in locating particles of a particular chemical composition or origin within a group of heterogeneous atmospheric or dust particles. An illustrative example is the determination of the sources of potentially toxic lead-containing particle in urban dusts using a particles electron microscope (11, 36). X-ray elemental maps of a dust subsample clearly show the preferential association of Pb with specific particles (Figure 6). In

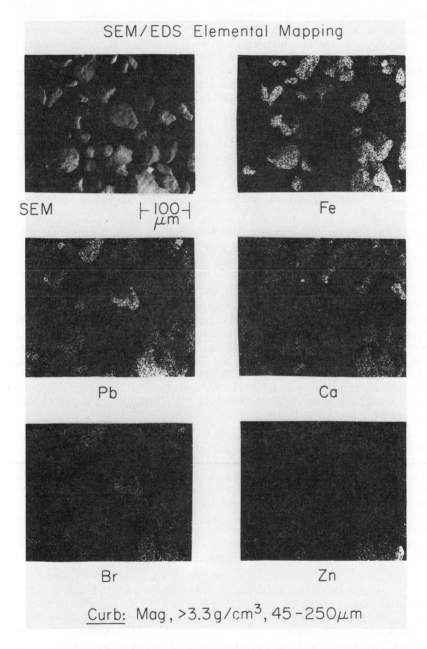

Figure 6. *X-ray elemental maps obtained using an electron microscope—urban dust sample containing automotive exhaust particles*

addition, the particles with the highest Pb content appear to be the most enriched in Br (Figure 6). Since the major forms of automotive Pb compounds include lead bromochlorides, it is apparent that the major lead-containing particles in the dust subsample are automotive in origin. Similarly, lead-containing particles from titanium-based paint are identified on the basis of the Pb and Ti maps, and the similarity in morphology and composition between the settled dust particles and authentic paint chips (Figures 7 and 8). A detailed discussion of these studies is published elsewhere (11, 36).

Similar elemental mapping capabilities are available using the Auger or ion microprobes. In the case of SAM, elemental maps reflect only the material present within about 20 Å of the particle surfaces, and their quality may be degraded by electrical charging effects. As a simple illustration of SAM elemental mapping capabilities, Figure 9 shows SAM images for a soot (C) particle produced during coal combustion and captured in a Cu microgrid (37). The particle lies between a grid opening and is virtually invisible in the Cu image, but shows up clearly in the C elemental map.

Elemental maps obtained using an ion microprobe will be highly surface specific as in SAM. However, since ion sputtering is destructive, repeated scans over the field of particles will penetrate deeper and deeper into the particle interiors. McHugh and Stevens have demonstrated the utility of IMP elemental maps in the identification and chemical characterization of oil soot particles in the atmosphere (38).

Surface Chemical Analysis. Electron spectroscopy of chemical analysis (ESCA) has been the most useful technique for the identification of chemical compounds present on the surface of a composite sample of atmospheric particles. The most prominent examples include the determination of the surface chemical states of S and N in aerosols, and the investigation of the catalytic role of soot in heterogeneous reactions involving gaseous SO_2, NO, or NH_3 (15, 39-41). It is apparent from these and other studies that most aerosol sulfur is in the form of sulfate, while most nitrogen is present as the ammonium ion. A substantial quantity of amine nitrogen also has been observed using ESCA (15, 39, 41).

The principal limitations of ESCA include the inability to detect elements present at trace concentrations within the analytical volume, and insufficient lateral resolution to characterize single micrometer-sized particles. The inability to characterize trace species is illustrated in Figure 10 for a sample of coal fly ash particles (11). The fly ash results from the noncombustible mineral components of the coal and consists largely of fused iron oxides and aluminosilicates (42). In addition, most elements are present in at least trace concentrations (22, 42), and many of these elements are highly enriched in the surface region of the particles (evidence for this will be discussed in the next section). However, the ESCA spectrum acquired over several hours of counting time indicates only the presence of detectable surface S and Ca in addition to the fly ash matrix constituents.

Precise measurements of ESCA binding energies suggest that the major oxidation states of the surface Fe, Al, Si, and S in fly ash are +3, +3, +4, and +6, respectively (43). The presence of an oxidized surface layer highly enriched in sulfate is of particular significance since it suggests that the surfaces of airborne fly ash may be important in controlling the heterogeneous oxidation of SO_2 to acid sulfates (11, 12).

Elemental Depth Profiling Analysis. An important aspect of a surface specific analysis is that it does not directly contrast the surface composition with that of the

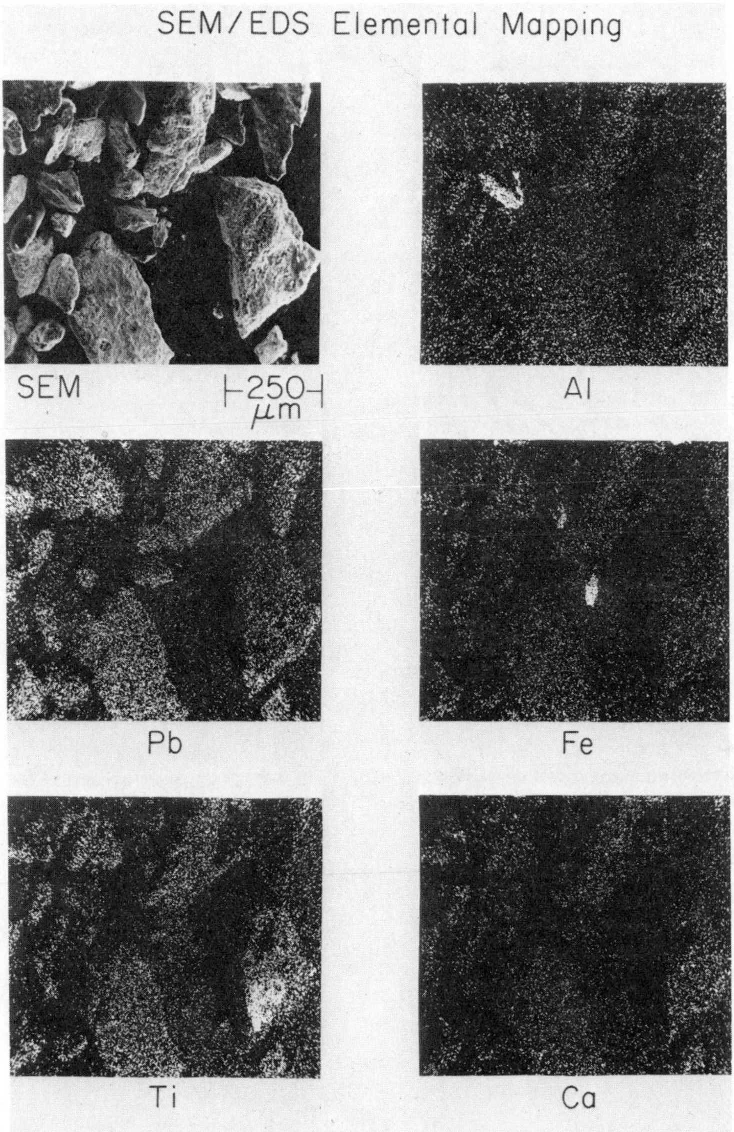

Figure 7. *X-ray elemental maps obtained using an electron microscope—urban dust sample containing paint chips with high lead content*

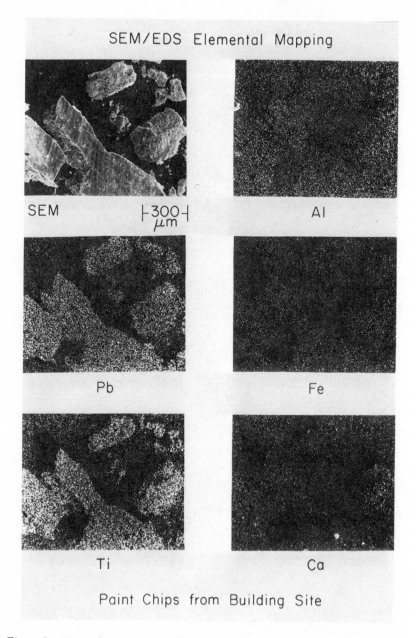

Figure 8. *X-ray elemental maps obtained using an electron microscope—paint chips removed from building adjacent to dust sampling site (Figure 7).*

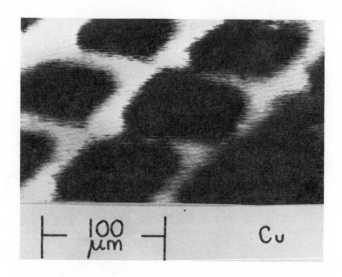

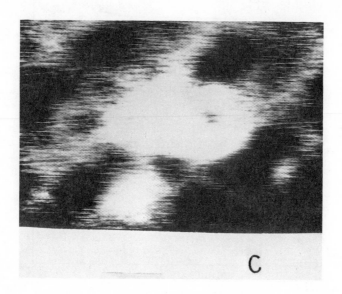

Figure 9.   *Elemental maps obtained using a scanning auger microscope—soot*
(C) particle trapped in a copper grid

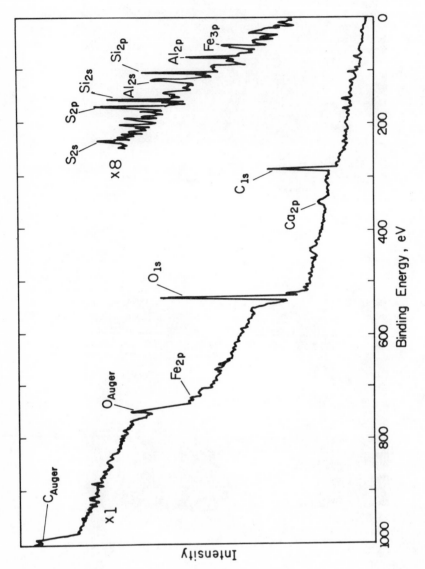

Figure 10.   ESCA spectrum of coal fly ash

sample's interior. The ion sputtering process previously discussed (Figure 4) can be used to remove successive layers of surface material, and thus enables one to perform elemental depth profiling (i.e., determination of relative concentrations as a function of depth). By monitoring the photoelectron, Auger electron, or secondary ion intensity as a function of sputtering time, one can obtain information on the relative concentration of an element as a function of depth into the sample interior. Principal limitations to this approach include the non-idealities of the ion sputtering process (11, 14, 30, 35), and the general difficulty in relating sputtering rates to an actual depth scale (25, 26, 30).

Certainly, the inherent lack of depth resolution of the EMP techniques minimizes the utility of sputter profiling combined with EMP analysis. However, gross comparisons of the "exterior" versus "interior" composition can be obtained by recording X-ray spectra before and after a minimum of several thousand angstroms of material are sputtered from the particle surface (13, 44). Similarly, ESCA is not very suitable when used in conjunction with sputter profiling for reasons that include: a) data acquisition rates are very slow, b) potential chemical information is lost since sputtering may alter the chemical forms of the elements present, and c) individual particles cannot be depth profiled (11, 14, 26).

The two instruments most suited to the sputter depth profiling of individual particles are the Auger and ion microprobes. The characterization of automobile exhaust particles produced in the combustion of leaded gasoline will be used to illustrate this point.

Automobile exhaust particles occur in two distinct morphologies as established by scanning electron microscopy (11, 36, 43, 45, 46). The first involves large particles ($>10\mu$m cross section) of irregular shape. Analysis using the EMP indicates that the major constituents include Fe, Pb, Br, Cl, and S as well as highly variable amounts of Ca and Si. Elemental depth profiles of the individual particles obtained using the Auger and ion microprobes indicate that Pb, Br, Cl, and S are surface predominant while Fe is not (Figure 11) (11, 12). The region of the surface enrichment is estimated to extend less than 1,000 Å into the particle interior (11).

The second characteristic particle type involves small ($<1\mu$m diameter) spherical particles containing mostly Pb, Br, and Cl with little Fe. Auger and ion microprobe analysis of such small particles are difficult, but preliminary studies suggest that none of the above elements are surface predominant.

The depth profiling studies suggest that two different processes govern the formation of automotive exhaust particles. The elemental surface predominance on large particles is attributed to the deposition of volatile Pb and S species (e.g. PbBrCl, SO_2) onto the surfaces of refractory iron-containing particles in the automotive exhaust system (11, 12). The iron-rich particles are probably derived from corrosion and ablation of the exhaust system. The smaller, more homogeneous particles may form by a nucleation process in which PbBrCl forms rather pure molten droplets when the exhaust system temperature falls below the saturation point (12).

The principal advantage of the ion microprobe (as opposed to the Auger microprobe) is the ability to obtain depth profiles for trace elemental species present in the analytical volume. The characterization of coal fly ash clearly illustrates this point (11-14). Auger detection limits are comparable to ESCA, and thus only elements with bulk concentrations greater than 1% by weight in fly ash (Si, Al, Fe, Ca, S, Na, K) can be

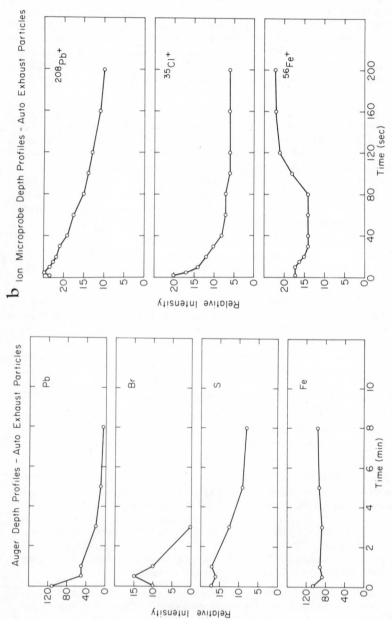

Figure 11. Auger and ion microprobe depth profiles of selected elements in auto exhaust particles: (a) auger; (b) ion microprobe.

characterized. Auger depth protiles indicate that K, Na, and S are enriched near the particle surfaces. On the other hand, ion microprobe mass spectrometry can be used to detect and depth profile trace species such as Pb and Tl with respective bulk concentrations of only 620 and 30 ppm by weight (Figure 12). Results of the IMP analyses indicate that a number of elements including Cr, K, Mn, Na, Pb, S, Tl, V, Zn are substantially surface enriched, whereas Al, Ca, Fe, Mg, Si, and Ti are not (11-14). Estimated average elemental concentrations in the outer 300 Å are compared to bulk concentrations in Table I (11). The above observations support the hypothesis that the more volatile elements and compounds are vaporized during combustion and then condense on the surfaces of co-entrained particles at lower temperatures (22).

The depth profile examples in Figure 12 also contrast the surface composition of fly ash before and after solvent leaching. The major purpose of the leaching is two-fold: a) to assess the solubility (i.e. potential environmental availability) of the surface enriched species, and b) to aid in the quantitation of surface region concentrations. In addition to Pb and Tl shown in Figure 12, the surface predominant region for other potentially toxic elements (e.g. Mn, Cr, S) is highly soluble, although fly ash is highly insoluble on a bulk basis (11, 14).

Solvent extraction or leaching also is used to dissolve organic compounds for subsequent trace characterization using chromatographic techniques. Not surprisingly, volatile organics (including carcinogenic compounds) are readily extracted from coal fly ash surfaces suggesting that a volatilization-condensation mechanism also governs the behavior of volatile organics (47).

Despite the very high surface specificity and sensitivity of the IMP, it is not very useful for the determination of surface organics on airborne particles. Reasons for this include the following: a) volatile organics may desorb in the high vacuum environment required for analysis, b) several hundred trace organic species may be present, and c) interpretation of molecular ion information is difficult even for the simplest of organic samples owing to the uncertainty regarding the ionization probability, gas phase stability, and fragmentation mechanisms for the molecular ions (48). Organic fragment ions such as C_2H_3+, or C_4H_7+ (Figure 13) do show a strong surface predominance in a fly ash sample, although the molecular precursors contributing these fragments cannot be identified (49), Other organic fragment ions with high stability, including aromatic species such as benzene and napthalene, also have been tentatively identified in the surface regions of fly ash using secondary ion mass spectrometry (50).

From the standpoint of environmental pollution, the surface analysis results discussed above have a number of ramifications:

1. Potentially toxic chemical species may be highly enriched in soluble forms on airborne particle surfaces. Such enrichments will have the greatest impact on the smallest, pulmonary-depositing particles with the highest surface area to volume ratios (11-14, 21, 22). Such a phenomenon is likely to occur for all particles produced by high temperature processes, both anthropogenic and natural (11-14).

2. The surface layer composition may influence the effectiveness of pollution control devices. For example, it is apparent that a surface region highly enriched is alkali-alkaline earth sulfates may enhance the fly ash particle collection efficiency of electrostatic precipitators (11, 12, 51-53).

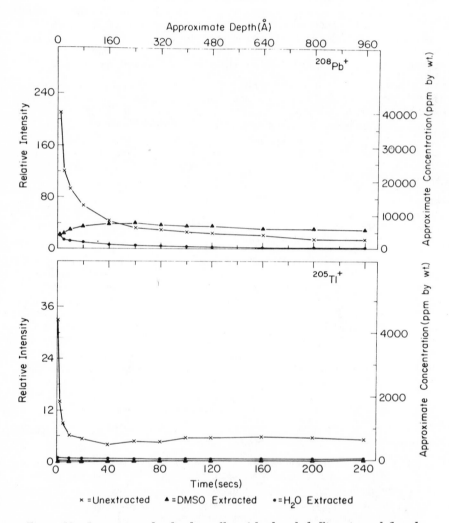

Figure 12. Ion microprobe depth profiles of lead and thallium in coal fly ash taken before (unextracted) and after solvent leaching with H₂O or DMSO

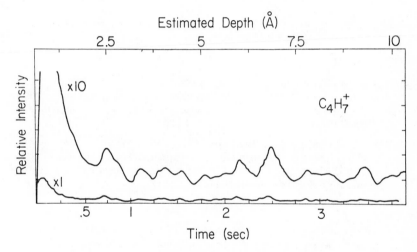

Figure 13. Ion microprobe depth profile of an organic fragment ion ($C_4H_7^+$) in coal fly ash

TABLE I. COMPARISON OF SURFACE REGION
AND BULK ELEMENTAL CONCENTRATIONS
IN COAL FLY ASH.

Element	Surface Region Concentration[a]	Bulk Concentration[b]
Pb	12,000	620
Tl	800	30
Cr	3,000	380
Mn	1,300	310
V	700	380
S	130,000	7,100
Fe	100,000	92,000
Ti	3,500	4,700
Mg	7,500	12,000

[a]Concentrations (ppm by wt) within ~ 300 Å of the surface estimated using relative elemental sensitivity factors (IMP) and solvent leaching studies.

[b]Bulk concentrations (ppm by wt) determined using spark source mass spectrometry.

3. The surface layer composition may effect catalytic activity. Surface enrichments of trace metals, for example, may enhance the catalytic role of particles in heterogeneous reactions in the atmosphere involving gaseous pollutants such as SO_2 (54, 55).

4. All of the surface effects listed above must be considered in the design of effective pollution control regulations. It is apparent that most current control devices may be poorly equipped to remove the particles with the greatest potential for deleterious environmental effects.

Conclusion

The chemistry of surfaces or interfaces is becoming increasingly important in a variety of scientific disciplines, including the environmental sciences. In the case of atmospheric particles, modern methods of surface chemical analysis are providing valuable insights into the chemical speciation, mechanisms of formation, sources, reactivity, and potential toxicity of pollutant species.

Acknowledgements

The assistance of the following individuals is gratefully acknowledged: Professor D. F. S. Natusch, Colorado State University, Fort Collins, Co; Professor C. A. Evans, Jr., and Dr. P. Williams, University of Illinois, Urbana, Il.

Support of this research was provided in part by National Science Foundation Grants DMR-76-01058, CHE-74-05745, ERT-74-24276; and the University of North Carolina School of Public Health and Department of Chemistry.

Literature Cited

1. Natusch, D. F. S., Bauer, C. F., Loh, A., in "Pollution Control", Vol. III, W. Strauss, ed., Wiley-Interscience, New York, N.Y., 1977, in press.

2. Zoller, W. H., Gladney, E. S., Duce, R. A., Science (1974), *183*, 198.

3. Hoffman, G. L., Duce, R. A., Hoffman, E. J., J. Geophys, Res. (1972), *77*, 5322 Geophys.

4. Duce, R. A., Zoller, W. H., Moyers, J. L., J. Geophys, Res. (1973), *78*, 7802 Geophys.

5. Blifford, I. H., Gillette, D. A., Atmos. Environ. (1972), *6*, 463.

6. Hirao, Y., Patterson, C. C., Science (1974), *184*, 989.

7. Gartrell, G., Jr., Friedlander, S. K., Atmos. Environ. (1975), *9*, 279.

8. Lee, R. E., Jr., von Lemden, D. J., J. Air Pollution Control Assoc. (1973), *23*, 853.

9. Miller, M. S., Friedlander, S. K., Hidy, G. M., J. Colloid Interface Sci. (1972), *39*, 165.

10. Luckey, T. D., Venugopal, B., Hutcheson, D., "Heavy Metal Toxicity, Safety, and Hormology", Academic Press, New York, N.Y., 1975.

11. Linton, R. W., "Physico-Chemical Characterization of Environmental Particles Using Surface Microanalytical Techniques", Ph.D. Thesis, University of Illinois, Urbana, Illinois, 1977.

12. Keyser, T. R., Natusch, D. F. S., Evans, C. A., Jr., Linton, R. W., Environ. Sci. Technol. (1978), *12*, in press.

13. Linton, R. W., Loh, A., Natusch, D. F. S., Evans, C. A., Jr., Williams, P., Science (1976), *191*, 852.

14. Linton, R. W., Williams, P., Evans, C. A., Jr., Natusch, D. F. S., Anal. Chem. (1977), *49*, 1514.

15. Novakov, T., Chang, S. G., Harker, A. B., Science (1974), *186*, 259.

16. Lusis, M. A., Wiebe, H. A., Atmos. Environ. (1976), *10*, 793.

17. Newman, L., Forrest, J., Manowitz, B., Atmos. Environ. (1975), *9*, 969.

18. Butcher, S. S., Charlson, R. J., "An Introduction to Air Chemistry", Academic Press, New York, N.Y., 1972.

19. Strauss, W., "Industrial Gas Cleaning", 2nd edition, Pergamon Press, Elmsford, N.Y., 1975.

20. Hatch, T. F., Gross, P., "Pulmonary Deposition and Retention of Inhaled Aerosols", Academic Press, New York, N.Y., 1964.

21. Natusch, D. F. S., Wallace, J. R., Science (1974), *186*, 695.

22. Davison, R. L., Natusch, D. F. S., Wallace, J. R., Evans, C. A., Environ. Sci. Technol. (1974), *8*, 1107.

23. Kaakinen, J. W., Jorden, R. M., Lawasani, M. H., West, R. E., Environ. Sci. Technol. (1975), *9*, 863.

24. Coles, D. G., Ragaini, R. C., Ondov, J. M., Environ. Sci. Technol. (1978), *12*, 442.

25. Czanderna, A. W., ed., "Methods of Surface Analysis", Elsevier, New York, N.Y., 1975.

26. Evans, C. A., Jr., Anal. Chem. (1975), *47*, 818A and 852A.

27. Birks, L. S., "Electron Probe Microanalysis", Wiley-Interscience, New York, N.Y., 1971.

28. Goldstein, J., Yakowitz, H., "Practical Scanning Electron Microscopy", Plenum Press, New York, N.Y., 1975.

29. Carlson, T. A., "Photoelectron and Auger Spectroscopy", Plenum Press, New York, N.Y., 1975.

30. Heinrich, K. F. J., Newbury, D. E., eds., "Secondary Ion Mass Spectrometry", National Bureau of Standards — Special Publication #427, Washington, D.C., 1975.

31. Bakale, D. K., Colby, B. N., Evans, C. A., Jr., Anal. Chem. (1975), *47*, 1532.

32. Cazaux, J. J., Microsc. Spectrosc. Electron. (1976), *1*, 73.

33. Hovland, C. T., Appl. Phys. Lett. (1977), *30*, 274.

34. Solomon, J. S., Meyers, V., Amer. Lab. (1976), *8* (3), 31.

35. Coburn, J. W., J. Vac. Sci. Technol. (1976), *13*, 1037.

36. Linton, R. W., Evans, C. A., Jr., Natusch, D. F. S., Solomon, R. L., Environ. Sci. Technol. (1978), *12*, in press.

37. Linton, R. W., unpublished results, 1977.

38. McHugh, J. A., Stevens, J. F., Anal. Chem. (1972), *44*, 2187.

39. Novakov, T., Mueller, P. K., Alcocer, A. E., Otvos, J. W., J. Colloid Interface Sci. (1972), *39*, 225.

40. Grieger, G. R., Amer. Lab. (1976), *8* (4), 77.

41. Chang, S. G., Novakov, T., Atmos. Environ. (1973), *9*, 495.

42. Natusch, D. F. S., Bauer, C. F., Matusiewicz, H., Evans, C. A., Jr., Baker, J., Loh, A., Linton, R. W., Hopke, P. K., Proceedings of the International Conference on Heavy Metals in the Environment, Vol. 2, Part 2, Toronto, Canada, 1977, p. 553.

43. Gordon, G. E., director, "Atmospheric Impact of Major Sources and Consumers of Energy", NSF-RANN Progress Report, University of Maryland, College Park, Md., 1975.

44. Loh, A., "Some Environmental Applications of Analytical Microscopy", Ph.D. Thesis, University of Illinois, Urbana, Illinois, 1975.

45. Boyer, K. W., Ph.D. Thesis, University of Illinois, Urbana, Illinois, 1973.

46. Olson, K. W., Skogerboe, R. K., Environ. Sci. Technol. (1975), *9*, 227.

47. Natusch, D. F. S., Environ. Health Perspect. (1978), in press.

48. Benninghoven, A., Surface Science (1973), *35*, 427.

49. Linton, R. W., Natusch, D. F. S., Williams, P., Evans, C. A., Jr., Paper #78, Tenth Annual Conference of the Microbeam Analysis Society, Las Vegas, Nevada, August, 1975.

50. Linton, R. W., Blattner, R. J., unpublished results, University of Illinois, 1976.

51. Bickelhaupt, R. E., J. Air Pollution Control Assoc. (1974), 24, 251.

52. Bickelhaupt, R. E., J. Air Pollution Control Assoc. (1975), 25, 148.

53. Kanowski, S., Coughlin, R. W., Environ. Sci. Technol. (1977), 11, 67.

54. Judeikis, H. S., Paper #80, Division of Environmental Chemistry, ACS National Meeting, New Orleans, La., March, 1977.

55. Fennelly, P. F., Am. Sci. (1976), 64, 46.

RECEIVED November 17, 1978.

Fourier Transform Infrared Analysis of Trace Gases in the Atmosphere

P. D. MAKER, H. NIKI, C. M. SAVAGE, and L. P. BREITENBACH

Research Staff, Ford Motor Company, Dearborn, MI 48121

Infrared absorption spectroscopy is a well recognized and powerful tool of the analytical chemist concerned with the identification and quantitation of the major components of liquid and solid samples. [1,2] Limitations arise in the detection of trace components due to interfering host absorptions and as well from the limited number of 'information channels' available. By this is meant the fact that typical samples have roughly ten absorption features, each on the order of 10 cm^{-1} wide, occurring in 2000 cm^{-1} wide interval of the infrared spectrum. The prospectus for applying the method to the analysis of gaseous atmospheric samples seems, in principle, highly promising. The major constituents, molecular oxygen and nitrogen, do not absorb infrared radiation, eliminating the problem of host interference. Further, the information content of the infrared absorption spectrum of a gas phase sample is vast in comparison to that of a liquid. Typical light molecular weight samples may have many hundreds of absorption lines, each only ~0.1 cm^{-1} wide. This latter fact, however, confounds the practical application of the technique. In order that Beer's Law apply, individual absorption features must be fully resolved. Without it, quantitation requires tedious, elaborate calibration for each unknown. Thus to realize the full potential of the technique, it is essential that the instrumental spectral bandpass be less than ~0.1 cm^{-1}. Conventional dispersive instruments capable of operating at this high resolution are found only in research laboratories and produce short segments of spectra (several hundreds of cm^{-1} long) at very slow rates. Thus gas sample analysis based upon dispersive instrumentation is time consuming and difficult, and not an attractive, general purpose method.

The advent of the modern commercial rapid scan, long stroke Fourier transform spectrometer [3,4] for the mid-infrared completely reverses the foregoing conclusion. In contrast to dispersive machines that isolate single infrared frequency intervals and measure them sequentially one at a time, the scanning interferometer effectively modulates each infrared wavelength λ at a characteristic frequency ($f = v/\lambda$, v being the moving mirror velocity), allow all wavelengths to reach the detector during the entire measurement period, and later determine the spectral intensity distribution via frequency analysis (Fourier transform) of the recorded signal. This multiplexing of the infrared signals results in an enormous signal to noise enhancement (known as Fellgett's advantage) that can reach $\sqrt{N}$ where N is the number of spectral elements contained in the record (N = 40,000 commonly)*. Further, the optical throughput of an interferometer is substantially greater than that of a dispersive instrument of equal

0-8412-0480-2/79/47-094-161$05.00/0

resolution, resulting in another signal to noise advantage (Jacquinot's). Sophisticated electronics amplify the infrared signal, digitally encode it, and transmit it to a dedicated minicomputer. There, the interferogram is Fourier analyzed to produce a recognizable, interpretable infrared spectrum. This spectrum has extremely accurate wavelength calibration, excellent signal to noise, and high photometric accuracy (provided adequate attention has been paid to the problems of detector saturation and ambient radiation). Further, the massive data sets that comprise each spectrum are stored in memory, ready to be manipulated by the analytical spectroscopist however he pleases. Altogether, the modern Fourier transform systems represent infrared spectroscopy in its finest hour. What follows here is a description of the distinguishing features of an FTIR spectrometer system assembled from components at the Ford Motor Company, Scientific Research Laboratory (5), and results illustrating the use of such systems in the routine analysis of trace contaminants in atmospheric samples.

Experimental Equipment and Procedure

In order to maintain complete control of both the optical arrangement and the data acquisition and processing, the decision was made to construct a Fourier transform spectrometer system entirely from individually selected components. The interferometer selected was the Eocom Model 7001, precursor of the current Nicolet and Carson-Alexiou machines. It is of the dual air bearing design, with concentric colinear infrared/HeNe laser optical paths. A germanium coated, potassium bromide beam splitter, together with a dielectric multilayer longwave pass filter, isolated the 500 - 3700 cm^{-1} spectral region for measurement. The optics begin with a simple water cooled glow bar source, and include a one meter White cell (Perkin Elmer) used at forty meter total path length as the main sample cell, and an auxiliary one-half meter double passed absorption cell used for taking reference spectra. The volumes of these cells are 68 and one liters, respectively. A 21.75 meter path length cell with 5.6 liter volume (Wilkes Scientific) is presently being installed. It promises half the sensitivity of the larger cell with 1/12 the volume. A three meter long, 186 meter path length, fifty liter volume cell has been constructed, its use awaiting the need for yet greater sensitivity.

A mercury-cadmium telluride detector (Santa Barbara Research Laboratory) used during the early stages of development limited the system long wavelength response to ~800 cm^{-1} and saturated badly. (6) A non-linear amplifier was designed and successfully employed to compensate the quadratic term in its response. A copper-doped geranium detector (also from SBRL) cooled by a closed cycle cryogenic system (Air Products and Chemicals) was installed to extend the long wavelength response and avoid the saturation problem, and is in current use. No significant non-linearity has been observed, but slight changes in the detector temperature cause noticeable changes in its responsivity when strongly absorbing samples reduce the infrared flux. This causes no serious problem in practice, merely offsetting calculated absorbance spectra by a constant amount. Minute motions of the detector imparted by the refrigeration system produce discernible noise bursts in the infrared electrical signal. It was found that this could be eliminated by simply disabling the compressor and/or second stage valving during data acquisition, but no significant reduction in overall noise power was observed, and the practice was not adopted. The optical components are supported on a granite table top and enclosed in an air-tight Lucite box. The spent air-bearing air, obtained from bottled nitrogen, purges this box and, for the most part, eliminates unwanted water vapor and carbon dioxide background absorption.

A fifteen bit analog to digital converter (Analogic) that required periodic adjustments to maintain linearity through zero, digitized the data. A PDP 11/40 CPU (Digital Equipment Corporation) with 28K core memory, dual RK05 magnetic disks for fast access mass storage, magnetic tape for data archiving, X-Y plotter, storage-display screen and Decwriter keyboard terminal comprise the computer system. All optical and electronic interconnections were handled in-house. The extensive interactive computer software, written in-house and mostly in assembly language, is under continuing development, new operational features being created to meet changing demands.

In the procedure eventually adopted, samples were collected from the various emission sources using a stainless steel bellows pump, transported to the instrument in Tedlar plastic bags, and interferograms recorded, all in a time as short as a few minutes or as long as several hours, depending on the distance to the source. No condensation or trapping of the gas samples was employed. In handling raw auto exhaust or other water-laden hot samples, predilution was employed, accomplished either by inflating the bag with diluent prior to introduction of the sample or with the aid of a constant-dilution sampler. (7) The usual problems associated with this grab-bag technique arise. Principal among them are sample plating to the bag walls, water condensation on the walls accompanied by the removal of soluble substances, and chemical reactions occurring within the bag and optical cell during processing. Experiments have shown that with quick processing, a good recovery is achieved for such gases as hydrogen cyanide, ammonia and sulphur dioxide, even in the presence of several percent water. Conversely, water itself has proven to be very difficult to handle via this simple procedure. Carefully dried sampling bags quickly adsorb water from the samples, and dry samples are quickly contaminated by water.

In order to assure full resolution of the pressure-broadened rotation-vibration line contours of the atmospheric pressure samples, the interferograms recorded are long enough to support digitization at $1/16$ cm^{-1} intervals (65,532 points per spectrum). As a compromise between total data taking time and eventual signal to noise, sixteen such interferograms are co-added before doing a Fourier transform. This data acquisition period lasts ~1.3 minutes. The ensuing transform requires an additional 3+ minutes, for a total data acquisition/processing time of <five minutes. The software to accomplish this employs a variation of the Cooley-Tukey algorithm (8), uses fixed-point double precision arithmetic, is written in machine language and uses double buffering between disk and core. The largest transform that can be handled within the confines of existing disk storage space is one of $2^{19} \approx .5$ million words. Using it, spectra of appropriately pressure-broadened gasses are obtained in ~30 minutes that demonstrate 0.05 cm^{-1} resolution with no serious apodization side lobes.

Results

Figure 1 illustrates typical results. Trace (A) shows the observed spectrum with empty sample cell, from 0 to 4000 wave-numbers. Absorption due to residual water vapor in the dry box is present. Other irregularities in what might be expected to be a smooth envelope are due to window deposits and to the multilayer dielectric thin-film filter used to block short wavelengths from the detector. Trace (B) shows a horizontal-scale expanded segment of this spectrum covering the range 1575-1600 cm^{-1}. The absorption lines are due to water, and are well resolved. Some ideal of the quality of the data is here apparent, but only after a vertical scale expansion of a factor of ten does noise begin to appear, as seen in Trace (C). While the two features at 1589 and

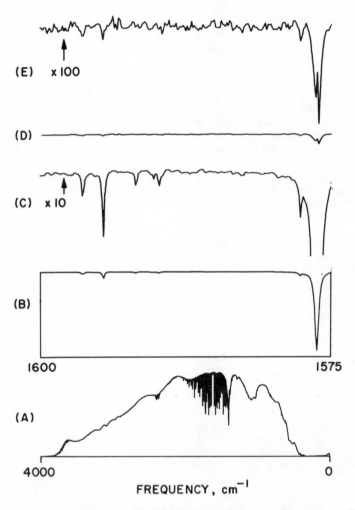

Figure 1.   Background (no sample) spectrum at different scale expansions illustrating system wavelength response, absorption attributable to typical amounts of residual water and carbon dioxide, and high signal-to-noise. The spectrum required 12 minutes of data acquisition at 8-cm optical retardation.

1591 cm^{-1} are only ~1.5% absorptions, much of the remaining ripple is not yet random noise, but is in fact due to yet weaker water vapor absorption and to a very weak window channel spectrum. This can be demonstrated by recording two spectra in succession and differencing them. Traces (D) and (E) result from such an experiment, (D) corresponding to segment (C) differenced between successive runs. Further gain expansion by 10X, as shown in Trace (E), is needed to reveal the true noise, now expanded by a total gain change of 100X from the original Trace (B). Imperfect water line cancellation is evident, indicating a slight (~1%) change in concentration between the two runs (taken ~1 hr. apart). As Trace (E) contains twice the noise power of a single record, the signal to peak-to-peak noise of (B) is seen to be ~1200:1. Certain factors essential to calculation of the photon shot noise are unknown, among them the source emissivity, detector quantum efficiency and the interferometer modulation efficiency. Assuming very reasonable values for them, the calculated shot noise limit is less than 1200:1. Thus it is reasonable to conclude that the system is operating in the photon shot noise limited regime. The result is atypical of usual procedure only to the extent that 144 interferograms were co-added prior to transformation. This represents ~12 minutes of data acquisition, nine times the usual length, and results in a 3:1 S/N enhancement. Thus, the customarily recorded spectra show only ~400:1 S/N. The results of Fig. 1 certainly prove that the S/N of the usually recorded spectra is not limited by rounding errors in the data acquisition or in the arithmetic.

A most difficult task for a Fourier transform instrument is the determination of the 0% transmission line and the concomitant question of absolute photometric accuracy. Checks of the system performance in this respect are made by recording the spectrum of concentrated samples that will produce absolute saturating absorption over short but continuous spectral intervals. Failure by the instrument to indicate 100% absorption in these regions can arise for a number of reasons. Among them are nonlinear detector or electronic response, analog to digital converter malfuction, inaccurate gain-ranging, etc. Absolute photometric accuracy of better than 0.5% over the entire 600-3600 cm^{-1} spectral range is achieved with our instrument.

Figure 2 serves to illustrate the remarkable reproducibility and precision of the spectra that are obtained. The upper trace shows the 950-1150 cm^{-1} region of the absorbance spectrum of 35 x 10^{-3} Torr (buffered to 700 Torr total pressure with ultra pure air and run at forty meters path length) of supposedly 90% pure ^{18}O methanol, as received from a supplier. The middle trace represents 12 x 10^{-3} Torr of ^{16}O methanol, as determined more than a year previously in the short auxiliary cell. The lower trace shows the difference. Note that the ^{16}O and ^{18}O spectra are completely distinctive, that the ^{18}O enrichment is in fact only ~65%, and that in the region of spectral overlap perfect separation of the two spectra can be achieved in spite of their great complexity. Further, the lower trace, obtained using archival reference data, represents a calibrated reference spectrum for ^{18}O methanol, even though no purified sample has ever existed.

Figure 3 is the absorbance spectrum of a sample of the ambient laboratory air drawn into the cell. Here, in accord with the usual procedure, the initially determined spectrum was first corrected for radiation that had reached the detector without having passed through the sample (room temperature background radiation entering the optical path via imperfect optical components and nonoverlap of the source and detector pupils and fields), ratioed against a zero-sample spectrum, and converted to absorbance. Trace (A) shows the spectrum from 3600-600 cm^{-1}. The massive absorbances seen here truncated at 1% transmission are due to water vapor and to carbon dioxide.

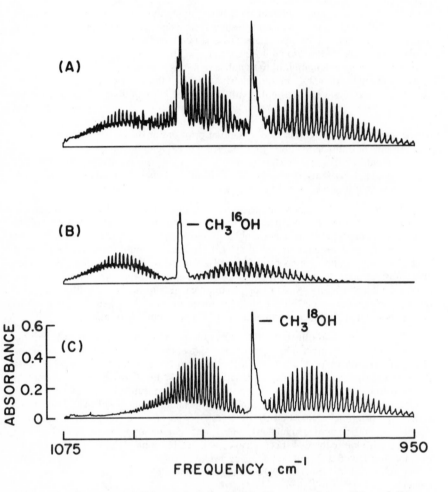

Figure 2. Spectrum of ^{18}O-enriched methanol, decomposed into ^{18}O and ^{16}O components

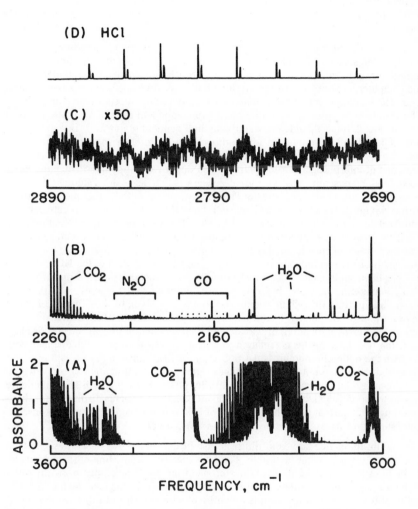

Figure 3. Absorbance spectrum of ambient laboratory air at different scale expansions together with a reference spectrum of hydrogen chloride (D)

The relative humidity thus measured was 32% and the carbon dioxide concentration was 360 ppm. Trade (B) is a blowup of the 2060-2260 cm^{-1} region, where absorption due to CO and N$_2$O can be recognized. Typical background concentrations of CO (0.28 ppm), N$_2$O (0.45 ppm) and as well CH$_4$ (1.6 ppm) were measured. In addition, a trace amount of HCl (0.12 ppm) was detected, as will be described in detail in subsequent paragraphs. This contaminant resulted from acid fluxed soldering being conducted in the room.

The procedure used to make these determinations, given the absorbance data, involved the successive subtraction of fractions of a spectrum of a carefully prepared sample of a pure compound. This subtractive procedure, carried out automatically by the computer, continues until the arc length of a residual spectrum has been minimized. The technique might be termed arc length determinant, subtractive desynthesis. The results of the procedure applied to a two component sample have already been seen in Fig. 2. It has the advantage of being independent of any base-line determination, it takes full advantage of all features of the reference spectrum (rather than concentrating on a single absorption line), and can be used with fair success in the presence of interfering absorptions from even unknown compounds. The method has been used extensively and with great success in our laboratory studies of various atmospheric chemical systems. (9-13) The ultimate power of the technique is illustrated by the two upper traces of Fig. 3. Trace (C) shows the room air spectrum between 2700 and 2900 cm^{-1} with the gain increased by 40 so that the noise is clearly visible. Trace (D) is the absorbance spectrum of 0.4 ppm HCl at the same gain. Visual inspection reveals no trace of correlated absorbance between the two. However, Fig. 4 shows the arc length of the residual spectrum as a function of the fraction of the HCl reference that has been subtracted. A clear minimum is seen, indicating the presence of 120 ppb HCl. Considering only the breadth of this minimum and the noise within it, the uncertainty of the value would seem to be about ±20 ppb. In further illustration, Fig. 5 shows a plot of the change in arc length and the optimum amount of the reference subtracted versus horizontal, or wavelength displacement of the two spectra of Fig. 6. Over the full range of this correlation plot (ignoring the datum at zero displacement) the average optimum amount is 0.8 ppb with an RMS uncertainty of 40.2 ppb. Our measured value, 120 ppb, is thus 3.0 times sigma which will happen once in twenty-two times in a purely statistical sample of fifty points (the size used). The data do not appear to be statistical, however. Since the arc length plot versus displacement shows its strongest feature at zero, the presence of HCl is undoubtedly real and a conservative estimate of its concentration would be 120±40 ppb.

While the above serves to illustrate use of the subtractive desynthesis method, the example is somewhat favorable. The HCl spectrum, consisting of strong, isolated single lines in a region of minimal interference from other molecules, is ideally suited to the purpose. This advantage is largely offset, however, by the fact that the signal-to-noise ratio in this spectral region is down by roughly a factor of three from its maximum.

Figure 6 illustrates another highly developed application of our FTIR facility. Trace (A) shows the complete absorbance spectrum, 600-3600 cm^{-1}, of a CVS bag sample of a typical auto exhaust. While dominated by water vapor and CO$_2$ absorption, several interesting features are visible. For instance, the carbon monoxide band is readily apparent, as is absorption due to heavy hydrocarbon (indicated by the broad unresolved C-H stretch band). In Trace (B) absorption due to CH$_4$, NO$_2$, formaldehyde and as well, water and heavy hydrocarbon can clearly be seen. In Trace (C),

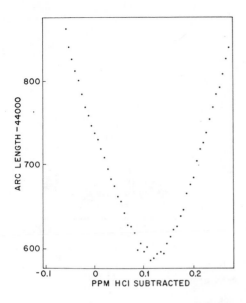

Figure 4.   Arc length of Figure 3C as a function of the amount of the hydrogen chloride spectrum, Figure 3D, that has been subtracted

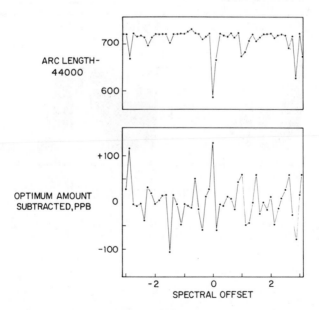

Figure 5.   Lower curve: amount of hydrogen chloride spectrum subtracted to minimize the arc length vs. the spectral offset in cm^{-1} between the two spectra shown in Figures 3C and 3D. Upper trace: achieved minimal arc length vs. offset.

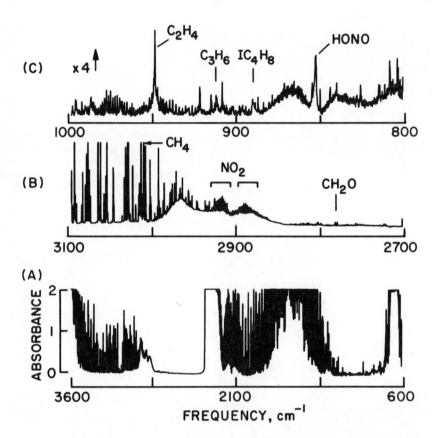

Figure 6.   Absorbance spectrum of typical cold–hot cycle, constant volume sampled auto exhaust, at various scale expansions with prominent bands identified

ethylene, propylene, isobutylene, nitrous acid and various water and CO_2 absorptions are visible. Software has been developed to automatically analyze spectra obtained from these bag samples. It integrates under carefully chosen sets of rotation-vibration lines (chosen to minimize overlap interferences with other known constituents), one set for each compound, using a locally determined zero. Residual interferences are measured by applying the analysis to spectra of the pure compounds. These are then removed by applying linear analysis, aided by matrix techniques. Table One presents the results of such an analysis applied to the spectrum of Fig. 6. Limits of error are substantially higher than were found in the analysis of the room air, due chiefly to the presence of so much water and carbon dioxide. As well, the method presumes that no compounds other than those listed will be present. Should a new gas enter the sample, it could cause either positive or negative interferences that would be propagated throughout the analysis by the interference correction procedure. The study further reveals that while dynamic range is signal-to-noise limited at about 100:1 (per absorption feature measured), the absolute concentration being measured can vary over much wider limits, provided only that one choose new absorption features of appropriate strength upon which to base the measurement.

This completely automated spectrum analysis procedure represents the final element in our effort to reduce to routine practice the quantitative analysis of similarly constituted gaseous samples by FTIR. It has seen wide and successful application within our laboratory, having been the principle analytic method for two extensive hydrocarbon species-specific auto exhaust catalyst efficiency studies, a comprehensive study of the gases emitted by passive-restraint air bag inflators, several controlled furnace atmosphere analyses, several stationary source stack emission checks and several health-related ambient atmosphere checks.

Returning briefly to Fig. 6(A), note the broad, unresolvable C-H stretching absorption. This feature arises not from a single compound, but comprises the integrated contributions of a large group of 'heavy' hydrocarbons, possibly even in the particulate state. The spectra of individual molecules in this class would also possess only broad featureless absorption contours, this because their mass is so large that individual rotation-vibration lines are packed more closely than the pressure broadened line width. The broad absorption features shift only slightly as the overall molecular weight increases, or even as the molecular geometry changes. This situation is further exemplied in Fig. 7. It shows the infrared absorbance spectrum of the effluent from an autobody paint bake-out oven (A), together with a smoothed version of the residual after the contributions from lightweight molecular species have been removed, (B). While such constituents as water, carbon dioxide, methane, carbon monoxide and formaldehyde can be quantitatively measured, the high molecular weight solvent and paint degradation/curing fractions, heralded by the intense but relatively featureless bands near 3000 cm^{-1} and 1100 cm^{-1} cannot even be unambiguously identified. The presence of molecular species containing CH, C = O, C-O-C and C-OH groups is certainly indicated, but even equipped with a multitude of reference spectra, including those of molecules known to be present by an independent mass-spec analysis, it proves impossible to give a completely satisfactory account of these features. While key spectra may still be missing, the inadequacy stems also from the fact that several trial combinations of spectra give equally appealing results. Thus the limitation of the technique is approached. As the molecular weight of the subject molecules grow, even though they are still in the gas phase, their spectra become less and less distinctive, their 'information content' lessens, and moreover, more and more candidates need to be considered. Ultimately, in spite of computer aids, an impasse is reached, just as in

TABLE I. Computer Report of the Quantitative Analysis of the Spectrum of Fig. 6

Compound	Amount	Err.[a], Unit	% THC
1 H_2O	1.51	0.1 %	0.000
2 CO_2	1.35	0.1 %	0.000
3 CO	781.29	10.0 ppm	0.000
4 HC[c]	112.00	8.0 ppmC[d]	0.693
5 NO	15.49	0.3 ppm	0.000
6 NO_2	47.25	0.3 ppm	0.000
7 N_2O	0.81	0.1 ppm	0.000
8 HONO	4.90	0.1 ppm	0.000
9 HCN	−0.12	0.2 ppm	0.000
10 NH_3	0.03	0.1 ppm	0.000
11 SO_2	0.19	0.2 ppm	0.000
12 CH_4	11.44	0.1 ppmC	0.071
13 C_2H_2	3.06	0.2 ppmC	0.019
14 C_2H_4	13.42	0.5 ppmC	0.083
15 C_2H_6	1.36	0.2 ppmC	0.008
16 C_3H_6	10.76	1.0 ppmC	0.067
17 IC_4H_8	6.08	1.0 ppmC	0.038
18 CH_2O	2.27	0.1 ppmC	0.014
19 HCOOH	1.00	0.1 ppmC	0.006
20 CH_3OH	0.28	0.1 ppmC	0.002

Total HC = 161.66 ppmC = 53.89 $ppmC_3$

Total NO_x = 67.65 ppm

(a) Estimated error of measurement
(b) Percent of total hydrocarbon
(c) Heavy hydrocarbon, representing the area under the unresolved CH stretching band, normalizing to raw fuel (indolene clear).
(d) ppmC = ppm÷molecular carbon number.
   Taken as 8 for HC.

* The signal to noise ratio is limited in any physical intensity measurement, however, by the statistical fluctuations in the photon flux (photon shot noise). This limit can be reached with Fourier transform infrared spectrometers.

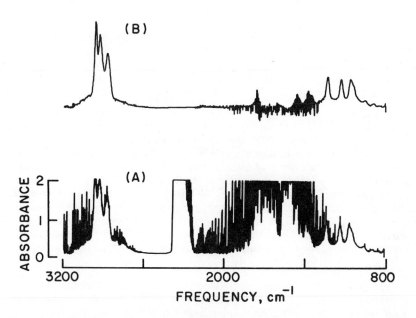

Figure 7.   Absorbance spectrum of the effluent from an auto body bake-out oven before (A) and after (B) the spectra of the low molecular weight fractions have been cancelled

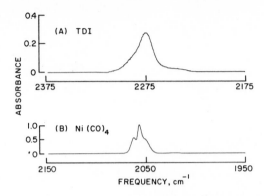

Figure 8. Segments of the absorbance spectra of toluene diisocyanate (A) and nickel carbonyl (B)

the case of the conventional IR analysis of liquid and solid samples.

Two final examples of the sensitivity and general applicability of the FTIR gas analysis technique are illustrated in Fig. 8. Trace (A) shows the spectrum obtained from an ultra-air filled 70 liter sampling bag into which had been injected, 18 hours previously, 4.8 microliters of TDI, toluene diisocyanate. On the basis of the single feature at 2273 cm^{-1}, it is estimated that ~50 ppb TDI could be detected. The lower Trace (B), shows the spectrum of nickel carbonyl. This highly toxic but unstable gas was found to decay rapidly at ppm concentrations in ultra air (50% lifetime ~15 minutes). Calibration of its spectrum was established by recording successive spectra at ten minute intervals and by attributing the increase in carbon monoxide concentration (calibration known) to an equivalent but four times slower decrease in nickel carbonyl concentration. The spectrum shown represents 0.6 ppm of the material. Note the extraordinary absorption strength. The detection limit is thus less than 10 ppb.

Conclusion

It is immediately clear that high resolution Fourier transform mid-infrared absorption spectroscopy is a powerful method for the analysis of trace constituents in the atmosphere. Most lightweight (molecular weight less than ~60) molecules can be detected at concentrations down to ~100 ppb in five liter samples. Complex mixtures can be handled, and trace components can usually be quantitated in the presence of huge concentrations of interfering substances. Once programmed, the computer aided analysis of qualitatively similar samples proceeds rapidly, the throughput rate reaching one sample every ten minutes, including printout. Sample preconcentration, drying, and carbon dioxide removal should extend the sensitivity by another order of magnitude. For ambient atmosphere analysis, extremely long absorption path lengths (1-5 kilometers) result in detectivities down to a few parts per billion. Limitations quickly arise as the molecular weight of the subject species rises. It is safe to say, however, that the broad application of gas phase analysis by FTIR is limited chiefly by the capital (>$125,000) and manpower (programmer as well as operator) investment required. It is expected that these costs will surely fall in the years ahead, and that ever greater reliance will be placed upon the technique. The future undoubtedly holds computerized data banks, centralized and administered by Federal-commercial agencies, much as the mass-spectroscopy data is now handled as described by S. R. Heller and G. W. Milne in this volume.

Literature Cited

1. See for instance: Alper, N. L., Keiser, W. E. and Szymanski, H. A., "IR — Theory and Practice of Infrared Spectroscopy", Plenum Press, New York (1970).

2. Avram, M. and Mateescu, G. H., "Infrared Spectroscopy, Application in Organic Chemistry", Wiley-Interscience, New York (1972).

3. See for instance: Bell, R. J., "Introductory Fourier Transform Spectroscopy", Academic Press, New York (1972).

4. Hanst, P. L., Lefohn, A. S. and Gay, B. W. Jr., Appl. Spec. (1973) 27, 188.

5. Maker, P. D., Niki, H., Savage, C. M. and Breitenbach, L. P., 30th Symposium on Molecular Structure and Spectroscopy (1975), Paper FA-10, Ohio State University, Columbus, Ohio.

6. Bertoli, F., Allen, R., Esteronitz, L. and Kruer, M., J. Appl. Phys. (1974) 45, 1250.

7. Butler, J., Ford Scientific Laboratory, private communication.

8. Bergland, G. D., Comm. A.C.M. (1968) 11, p. 703.

9. Niki, H., Maker, P. D., Savage, C. M. and Breitenbach, L. P., Chem. Phys. Lett. (1977) 45, 564.

10. Niki, H. et al., Chem. Phys. Lett. (1977) 46, 327.

11. Niki, H. et al., Anal. Chem. (1977) 49, 1346.

12. Niki, H. et al., J. Phys. Chem. (1978) 82, 132 and 135.

13. Niki, H. et al., Chem. Phys. Lett. (1978) 55, 289.

RECEIVED November 17, 1978.

Opto-Acoustic Spectroscopy Applied to the Detection of Gaseous Pollutants

C. K. N. PATEL

Bell Laboratories, Murray Hill, NJ 07974

INTRODUCTION

Use of spectroscopy to study molecular energy levels is a commonly accepted technique.[1] Once absorption coefficients and the absorption frequencies are known, quantitative determination of a given molecular specie in a gaseous sample is easy. And thus spectroscopic techniques have been used for a very long time for pollution detection. Both dispersive as well as nondispersive sources have been used. A simple measurement scheme (see Fig. 1) would involve a radiation source capable of providing radiation at the frequency of absorption of the molecular specie to be detected, an absorption cell for containing the gaseous sample and two detectors for detecting the radiation that enters and emerges from the cell. Typically, most molecules have their fundamental vibrational-rotational frequencies in the infrared region from $\sim 2 \ \mu m$ to $\sim 15 \ \mu m$. And thus the source of tunable/fixed frequency radiation has been a black body radiation that is filtered either through a spectrometer or a set of filters. By tuning the source frequency one can obtain a "finger-print" absorption which can be used for identifying and for quantitative determination of the molecules (see Fig. 2(a)). The "finger-print" absorption can also be obtained in at least two other and different ways. The first is the technique of modulation spectroscopy where the frequency of incident radiation has a periodic variation given by

$$\nu(t) = \nu_o + \delta\nu \sin \omega_m t \tag{1}$$

In this situation one chooses $\delta\nu$ to be a small fraction of the linewidth of the molecular transitions that is being measured. Instead of measuring the absolute level of the transmitted radiation (as was done in the first technique), we can measure the ac component at ω_m which arises due to the frequency dependent absorption in the absorber. This technique, called the modulation technique, yields a differentiated signal corresponding to the absorption as seen in Fig. 2(b). Finally, there is one more technique that is used for determining the absorption in a sample. This is the calorimetric technique which relies on the fact that in our absorption measurement, the fractional radiation power that does not emerge from the absorption cell must have remained in the sample itself (assuming for the moment that the windows of the cell are totally transparent). In a gaseous sample the power absorbed by the gas is initially in the form of vibrational excitation of the molecules. Before we touch upon the methods of measurements it is necessary to see the de-excitation pathways through which the vibrationally excited molecules return to ground states. We have

0-8412-0480-2/79/47-094-177$05.00/0

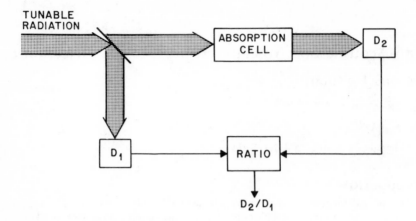

Figure 1.   Sketch of absorption measurement set up.

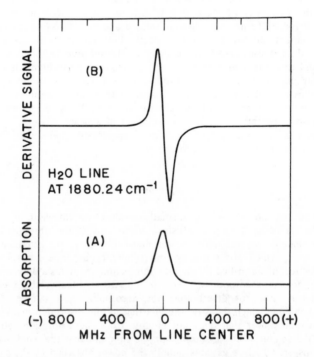

Figure 2.   (a) Absorption signal on a H_2O line; (b) derivative signal obtained
using modulation spectroscopy

$$M + h\nu \text{ (resonant)} \rightarrow M^* \text{ (excitation)} \tag{2}$$

$$M^* \xrightarrow{k_{em}} M + h\nu \text{ (radiative relaxation)} \tag{3}$$

$$M^* + X \xrightarrow{k_{v \rightarrow t}} M + X + kT \text{ (nonradiative relaxation)} \tag{4}$$

as the principal excitation/de-excitation mechanisms. If the vibrationally excited molecules relax primarily through the collisional deactivation described in Eq. (4) where X is another molecule or wall, it is easy to see that temperature of the gas will rise. Thus, for calorimeteric detection of absorption the nonradiative decay process appears to be necessary. In fact, what is necessary is that the vibrational to translational conversion rate $k_{v \rightarrow t}$ be substantially faster than the radiative relaxation rate, k_{em}. In the infrared region for many molecular vibrational-rotational transitions, the spontaneous transition probability A is small (because of its ν^3 dependence for similar dipole matrix element connecting two states) compared to $k_{v \rightarrow t}$ at pressures of $\geqslant 1$ Torr. Thus the condition stated above is automatically satisfied. Even at lower gas pressures or for those molecules where A coefficients are known to be large, a careful look will convince one that the condition stated above is much too restrictive. This can be seen from considering a totally unfavorable situation of $A \gg k_{v \rightarrow t}$. Naively it may seem that all the excitationl energy will be lost through reradiation. However, we should remember that the reradiation will occur in all directions (4π steradian). In practical situations absorption cells are cylindrical (see Fig. 1 for example) with a small diameter/length ratio. Thus only a small fraction of the spontaneously re-emitted radiation will escape without bouncing off the cell walls. Thus if the cell walls are made even nominally nonreflective, a very large fraction of the spontaneously re-emitted radiation will cause heating of the cell wall and consequently heating of the gas. While there is some loss in efficiency of conversion of vibrational energy of the molecules to thermal energy, proper cell design can minimize the differences between situations where we have $A \gg k_{v \rightarrow t}$ or $A \ll k_{v \rightarrow t}$. In a typical situation in the infrared, then, it is reasonable to expect that sizable if not all of the vibrational energy of the molecules excited through Eq. (2) will end up in thermal energy of the gas. There is only one exception to this and that has to do with some unique situations such as self-induced transparency in which there is forward re-emission of the radiation (along the well defined direction of initial laser excitation). The fact that radiative relaxation need not result in a reduced efficiency for calorimetric detection scheme if the cell walls are absorbing (i.e., nonreflecting) is well recognized in construction of Golay cells for power measurements. Having converted the molecular vibrational energy into thermal energy, we are still left with the necessity of measuring the change in the gas temperature. This is best done, in a relatively closed gas cell, by measuring the gas pressure change that accompanies the temperature change by using a sensitive acoustic microphone.[2] This scheme, the opto-acoustic detection, was first described[3] in 1880 and has been used in the intervening years only sporadically.

Before we proceed any further, it is important to compare the above three techniques for their ability to measure small absorption coefficients - an ability which makes spectroscopic studies easier and pollution detection possible. The straightforward scheme, shown in Fig. 1, has been shown to be capable of measuring absorptions as small as 10^{-4}. The limitation here is set by amplitude fluctuations in incident tunable radiation, as long as the intensity (incident and/or transmitted) is $\geq 10^{-4}$ NEP of the detectors. The modulation spectroscopy scheme gets around the problem arising from amplitude fluctuations and can be used for measuring absorptions as small as 10^{-8} assuming that the ac component of transmitted power induced by a frequency dependent absorption of the gas is larger than NEP of the detector. Since this ac component is proportional to both the gaseous absorption as well as incident intensity, I_{in}, the sensitivity or the ability of the modulation technique to measure small absorptions improves as $1/I_{in}$ up to the 10^{-8} limit quoted above. Beyond this level, because the shot noise generated in the detector by the large background radiation (i.e., unabsorbed radiation) little improvement is possible. The opto-acoustic technique, finally, measures a quantity that is proportional to the product of the incident radiation intensity and the gaseous absorption. Thus the ability to measure small absorption coefficient with the opto-acoustic technique improves $1/I_{in}$ and there is no fundamental limit as to how low an absorption that can be measured if arbitrarily high incident powers are available. The limitation on the minimum detectable absorption is described by

$$P_{min} \text{ (transducer)} \leqslant I_{in}(1-e^{-\alpha l}) \tag{5}$$

where P_{min} (transducer) is the minimum power detection capability of the transducer (in the present case an acoustic microphone), α is the absorption coefficient to be determined and l is the gas cell path length. The P_{min} (transducer) is the effective NEP of the system taking into account the conversion of absorbed power into a minimum detectable pressure change in the gas which depends on a variety of factors including the microphone response, geometry of the opto-acoustic cell, etc. Since P_{min} (transducer) is a predetermined quantity, Eq. (5) clearly shows that minimum measurable absorption coefficient, α_{min}, varies as $1/I_{in}$. The only fundamental limitation on measuring smaller and smaller α_{min} by increasing I_{in} is the possible spontaneous or stimulated Brillouin or Raman scattering by the majority fraction of the gas sample. Of course, there are other limitations which have to do with the design of the opto-acoustic absorption. For example, the absorption of incident radiation in the input and exit windows will give rise to a background signal which is independent of wavelength tuning. In some sense this background signal corresponds to background signal in case of the modulated absorption measurement technique discussed above. There is a difference, however, in that by proper choice of window material, the window absorption can be made as small as $10^{-6} - 10^{-7}$. Further rejection against the gas pressure signals originating in the windows for the ultimate small absorption measurement in the gas can be obtained through discriminating between volume generated radial pressure waves arising from uniform absorption in the gas and the longitudinal pressure waves generated by windows. Two relatively straightforward techniques have been devised for such a discrimination. The first one is to make the opto-acoutic cell resonant[4] in a radial acoustic mode at the frequency (acoustic) at which the incident radiation is chopped. This has the result of resonant enhancement of the acoustic signal generated in the volume of the gas, leaving the window signal unaffected. The second scheme involves introducing appropriate acoustic baffles in the optical path length between the windows and the bulk of the absorption cell. Such baffles are easily installed since usually tunable (either continuous or step-wise) lasers are used as sources of radiation (see below) and the radiation is usually focused through the opto-acoustic cell. Of course, the possibility of having window-less opto-acoustic cell is always there which then removes all possible instrumental limitations on the minimum absorption that can be measured with the opto-acoustic technique keeping in mind that any incident radiation that directly impinges upon the cell wall will give rise to large background signal. Another point that needs to be made here is that with proper care to maintain the OA cell pressure constant, the sensitivity of detection of a particular molecular specie remains unchanged when SFR laser power normalization is taken into account. This long term constancy of calibration is an important parameter in the overall usefulness of the technique under discussion. Since the technique being used is primarily a spectroscopic one, it can be easily extended to almost any of the pollutants by proper choice of a tunable laser frequency.

TUNABLE SOURCES

As seen above, high incident powers are necessary in order to take advantage of the unique capabilities of the opto-acoustic detection technique. It is for this reason that the opto-acoustic detection schemes have seen the great upsurge only with the availability of a variety of high power continuously tunable and fixed frequency step tunable lasers starting in 1964 with the discovery of the CO_2 laser.[5] Gaseous spectroscopy with the opto-acoustic measurement technique has been carried out primarily with either the step tunable CO_2 or the CO (Ref. 1) lasers or with the magnetically tunable spin flip Raman (SFR) lasers.[7] It is not the intent of this article to dwell upon the characteristics of these sources and thus I will leave it to the readers to acquaint themselves with the operating principles, characteristics, and techniques of these lasers. The CO_2 laser (with its various isotopic species) is step tunable from $\sim 9\ \mu$m to $\sim 11\ \mu$m, the CO laser is step tunable from $\sim 4.8\ \mu$m to $\geq 7.0\ \mu$m and the SFR laser is continuously tunable from $\sim 5.2\ \mu$m to $\sim 6.5\ \mu$m and from $\sim 9\ \mu$m to $\geq 17\ \mu$m depending upon the fixed frequency pump laser that is used.

OPTO-ACOUSTIC ABSORPTION CELL

A highly successful improvement in the opto-acoustic cell design occurred in 1977 when special electret microphones with built-in FET preamplifiers were substituted for an earlier cylindrical OA cell geometry.[8] Figure 3 shows these two designs. With the new scheme, we have been able to measure an absorption coefficient of $\leq 10^{-10}$ cm^{-1} corresponding to a measurement of fractional absorption of 10^{-9} of the incident radiation in a 10 cm long OA cell when a tunable SFR laser power of ~ 100 mW was used. The above sensitivity corresponds to a measurement of absorbed power of $\sim 10^{-10}$ W cm^{-1} for an s/n ratio of ~ 1 in a 1 sec integration time. The sensitivity was also checked by careful measurements of controlled direct heat input into the cell by using a heater wire along with center of the cell and passing an ac electrical current through the wire.

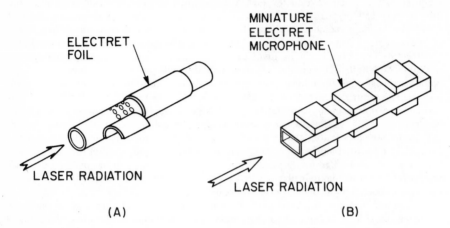

Figure 3.   OA cell construction

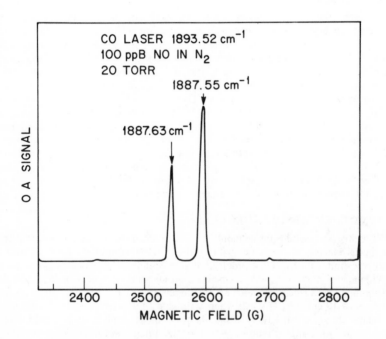

Figure 4.   OA signal for absorption from NO lines for a 100 ppb NO sample in nitrogen at a total pressure of 20 Torr vs. SFR laser magnetic field

The measurements of minimum detectable absorption were carried out by studying the s/n ratio in experiments to opto-acoustically analyze a 100 ppB NO sample in nitrogen at a pressure of 20 Torr. An opto-acoustic signal spectrum vs. the SFR laser magnetic field is shown in Fig. 4. From these studies, we conclude that at a total gas pressure of 20 Torr in the OA cell, using a tunable SFR laser power of $\sim$100 mW we can detect $\sim$10^7 mol/cm^3 of NO with an s/n ratio of 1 with an integration time of 1 sec. Such a detection capability, unique for an infrared spectroscopic system corresponds to a volumetric mixing ratio of $\sim$1:10^{12} at an atmospheric pressure. The variation in the sensitivity of the OA cell with a change in the gas pressure is discussed in Ref. 8 where we see that the OA cell sensitivity is independent of the gas pressure for pressure greater than 20 Torr. This characteristic has desirable impact on the line or step tunable CO$_2$ and CO lasers for pollution detection sources as we shall see a bit later. At the same time, the ability to use the OA cell at lower gas pressures (with a reduced sensitivity, though) has been proven to be significantly important in high resolution spectoscopy of ground state as well as excited states of molecules.[7]

Before I move on to the description of specific pollution detection experiments, some discussion of the dynamic range of the opto-acoustic absorption cell mentioned above is due. At the low absorption end, it is clear that the limitation is set by the equivalent NEP of the OA cell and the available laser power. At the present, the low end limit is $\sim$10^{-9} cm^{-1}. At the high end, two possible limits exist. The first one is the saturation of the signal arising from the microphones as the "sound level" in the OA cell increases with increasing absorption. However, this is not a fundamental limitation since the acoustic power generated in the OA cell can be easily reduced by reducing the laser power. The second limitation is a more serious one and of a fundamental nature. This has to do with the absorption coefficient getting to be so large that a substantial fraction of the input laser radiation is absorbed in the OA cell. It is clear to see the reason behind this limitation. The acoustic power generated can not possibly exceed the optical input power. In fact, when the acoustic power begins to approach 10% of the optical power, the response of the OA cell must begin to saturate. Thus the upper limit for linear response of the OA cell is set at $\sim$10^{-1} cm^{-1}, implying a dynamic range for the OA absorption measurements of eight orders of magnitude. Experimentally, we have verified the linearity of the response of the SFR laser based OA spectrometer over about eight orders of magnitude for detection of NO. This unique dynamic range allows the same instrumentation to be used for ambient (very low level) trace constituent detection as well as for source emission detection (very high level).

POLLUTION DETECTION WITH OPTO-ACOUSTIC SPECTROSCOPY

As mentioned above, pollution detection using opto-acoustic spectroscopy has utilized both continuously tunable SFR lasers as well as step tunable CO and CO$_2$ lasers. Under the most ideal circumstances, the tunable SFR laser is properly suited for pollution measurements since the continuous tunability allows one to obtain a complete "finger-print" absorption of the specie of interest and identify as well as quantitatively measure its concentration in the presence of other gaseous constituents at much higher concentrations. The step tunable CO or CO$_2$ lasers, on the other hand, offer the simplicity of their being primary laser sources, the ease of their operation and their very high power outputs. In the following I will deal with OA pollution detection measurement technique using the tunable SFR lasers and the step tunable CO or CO$_2$ lasers separately.

SFR LASER - OA ABSORPTION POLLUTION DETECTION

Figure 5 shows an experimental setup for measuring gaseous pollutant concentration using an SFR laser-OA absorption measurement system. The primary data output from the system is the opto-acoustic absorption signal as a function of the magnetic field used for tuning the SFR laser. The magnetic field values are eventually converted to SFR laser frequency, ω_{SFR}. Using the tuning relation

$$\omega_{SFR} = \omega_{pump} - g\mu_B B \qquad (6)$$

where ω_{pump} is the pump laser frequency, g is the g value of the electrons in the SFR laser crystal, μ_B is Bohr magneton and B is the magnetic field. Unfortunately, g is a slowly decreasing function of B and thus tuning of the frequency is less than linear in B. However, detailed measurements exist which allow accurate conversion of known B to the SFR laser frequency.

Ambient and Source Level Pollution Detection

One of the first uses of tunable laser based OA technique for pollution detection[9] was reported in 1971. This consisted of ambient (under a variety of situations) as well as source level (automobile exhaust, in the present situation) measurements of nitric oxide. Figure 6(a) shows a 20 ppM NO in N_2 calibration opto-acoustic spectrum. The lines marked with arrows arise from absorption due to NO while remaining ones are H_2O absorption lines arising from the residual H_2O in the nominally dry calibration sample. Figure 6(b) shows ambient air analyzed in a fairly active parking lot. From the strength of the NO absorption #8, which shows the least interference from H_2O, the nitric oxide concentration in the parking lot is estimated to be ~ 2 ppM. Figure 6(c) shows an analysis of automobile exhaust gases. (Note that the sensitivity scale has been changed.) The particular automobile exhaust is seen to contain $\sim 80-100$ ppM of NO. Because of the enormous signal/noise ratio for NO detection available in SFR laser - OA detection technique, the measurements shown in Figs. 6(b) and 6(c) are carried out in real time with typical time for each measurement of less than 1/10 second.

Detection of Minor Constituents in the Stratosphere

The ultralow level detection capability described above has found an important application to the real time measurement of minor constituents in the stratosphere, whose chemistry as well as composition especially regarding the trace gases has been, until very recently, only subjects of extensive theoretical studies and model making. The nitric oxide catalytic destruction of stratospheric ozone has been subject to some controversy especially with the expanding commercial SST flights in the stratosphere. Without going through the details (for which see Ref. 10) NO/NO_2 are thought to play a crucial role in the ozone balance. Expected NO/NO_2 concentrations are $\sim 10^9$ mol/cm^3 and known ozone concentration (at ~ 28 kM) is $\sim 10^{12}$ mol/cm^3. The following reactions summarize NO/NO_2-O_3 interactions (only primary ones):

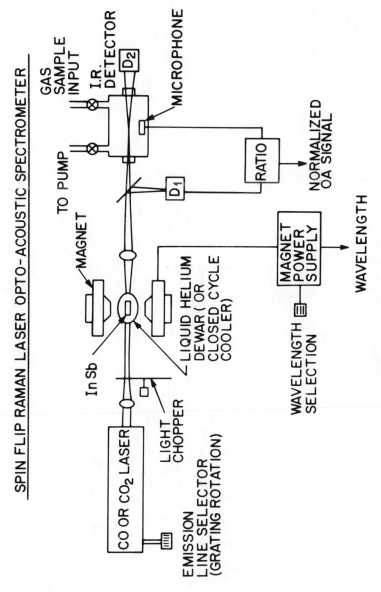

Figure 5.   SFR laser–OA pollution detection spectrometer

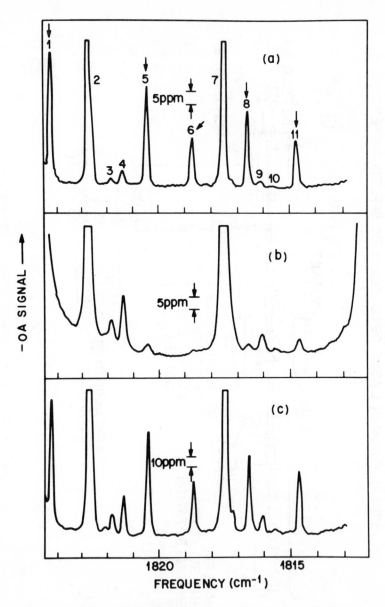

Figure 6.   (a) OA signal vs. B for 20 ppM NO calibration sample; (b) OA signal vs. B for ambient air in an active parking lot; (c) OA signal vs. B for automobile exhaust gases

$$NO_2 \overset{h\nu}{\rightarrow} NO+O \tag{7}$$

$$NO_2+O \rightarrow NO+O_2 \tag{8}$$

$$O_3+NO \rightarrow NO_2+O_2 \tag{9}$$

These reactions emphasize the importance of solar radiation in NO/NO_2 catalytic destruction cycle of ozone. One can immediately see that to provide any reliable observational basis for importance of NO/NO_2 in ozone balance, we must have not only NO/NO_2 concentration but also its diurnal variation which provides proper check on the time constants for the reactions described in Eqs. (7)-(9).

To provide the necessary data on NO, the entire SFR laser - OA absorption system was enclosed in a capsule and flown to an altitude of 28 kM for real time in situ measurements.[11] Figures 7(a) and (b) show the OA spectrum of ambient air (at 28 kM) analyzed before sunrise and at local noon. The lack of NO before sunrise and large concentration of NO at room is clearly seen. Figure 8 shows a summary of all the data compiled on two such balloon flights. We see that the measurements provide 1) absolute concentration of NO and 2) its diurnal variation. Many of the details of the model proposed above are confirmed (see Ref. 11 for details).

Recent discussions of stratospheric chemistry have dealt with the effect of freons[12] on ozone balance through a Cl/ClO catalytic destruction of ozone. The fundamental absorption band of ClO is measured to be at ~ 11 μm. Isotopically substituted CO_2 laser based OA absorption measurement technique should allow us to carry out fundamental measurements on ClO and its diurnal variation in the stratosphere to provide yet another important parameter (in addition to NO above) in the stratospheric ozone chemistry.

STEP TUNABLE LASER - OA ABSORPTION POLLUTION DETECTION

It is easy to see how a spectroscopic technique that uses a continuously tunable source (laser) can provide unique determination of a given molecular species. Both CO_2 and CO lasers provide a multitude of step tunable (by incorporating a grating with the laser cavity) laser lines in the 10 μm and the 5 μm regions, respectively. At low gas pressures, where the molecular absorption lines are essentially Doppler broadened, it is easy to convince ourselves that an exact coincidence between any of the step tunable laser lines and an absorption line of a given molecular specie will only be accidental and thus in general step tunable laser - OA absorption technique may be of no practical significance. However, pressure broadening of absorption lines of molecules can be advantageously used for obtaining absorption of laser lines which are close to a given absorption time but which are not exactly coincident. The trouble, however, arises from the facts that in general we may not be able to obtain entire

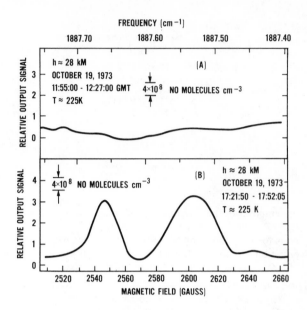

Figure 7. (a) OA signal vs. B at 28 kM before sunrise; (b) OA signal vs. B at 28 kM at noon

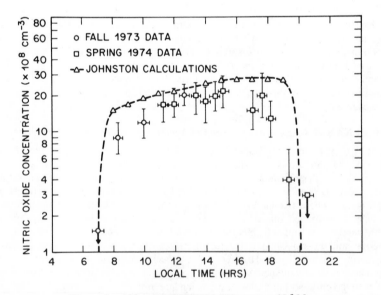

Figure 8. NO concentration vs. time at ~ 28 kM

"finger-print" pattern of absorption of the molecule and that absorption from another molecule may also give rise to an interference signal when pressure broadening is a necessary requirement for the success of the technique. (The overriding reason for the use of step tunable fixed lasers, as opposed to the SFR lasers described in the previous section, arises from the tremendous possible simplification of the experimental apparatus.) The interference problem is largely solved by making measurements, successively, using several different step tuned laser lines, creating a matrix of absorption signals obtained from the OA cell and then with the use of an on-line computer derive concentration of each one of the gaseous constituents making use of a calibration matrix obtained by using known calibration gas samples. Figures 9(a) and (b) show the OA absorption signals from 1000 ppM NO_2 and 1000 ppM NH_3 as a function of step tuned wavelength of a CO laser. These figures point out the importance of the proper choice of CO laser lines for detection of NO_2 and NH_3 with minimum interference from each other. We have shown earlier[13] that his technique provides capability of ppB level detection for many of the common industrial pollutants.

An important application of the step tunable CO_2/CO laser - OA absorption pollution measurement technique has been described recently.[14] We have studied the effluent from a catalyst, nominally designed for oxidation of CO to CO_2. Such a catalyst was tested under reducing conditions with inlet gases which reflect typical concentrations of CO, NO, H_2O etc., from untreated automobile exhaust. Analysis showed that sizable conversion of NO to HCN and to NH_3 takes place under the reducing conditions and pointed to some possible problems associated with the catalytic treatment of automotive exhaust. The HCN measurements carried out with the step tunable CO laser - OA spectrometer showed that we are able to accurately determine sub-ppM concentration of HCN in presence of very large concentrations of interfering gases such as NO_2, H_2O, etc. Tables Ia and Ib give the laser lines used and the sensitivities for the detection of HCN etc. These data are obtained by rapid scan of the CO laser frequencies under computer control[15] (Data General Eclipse) which allows continuous optimization of operating conditions and data collection as well as data manipulation for direct quantitative determination of various pollutant gases. The experimental setup is shown in Fig. 10. As opposed to gas chromatography techniques, we are able to provide transient measurements which are important in the analysis of the reactions as well as in the study of time dependent poisoning of the catalysts.[16] Figure 11 shows HCN concentrations in the effluent as a function of the catalyst temperature for three different catalysts. Figure 12 shows transient studies of HCN and NH_3 with effects of the addition of oxygen to the inlet gases. These studies have been carried out to include the effects of various additives to the inlet gases. Recent studies[17] of actual automobile catalysts under operating conditions substantiate our original findings where the step tunable CO laser - OA spectrometer played a crucial role.

CONCLUSION

In the above I have very briefly sketched the opto-acoustic pollution measurement techniques and described a few examples where the technique has been applied to problems of scientific, industrial, and societal significance. It is clear that these techniques are simple and practical and will play increasing by important roles in pollution monitoring.

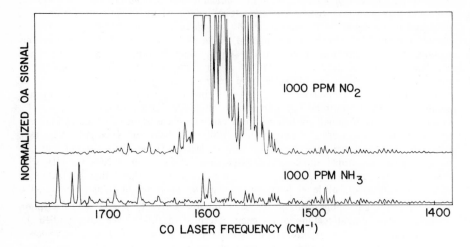

Figure 9. OA signal vs. step-tuned CO laser wavelength for 100 ppM NO₂ in nitrogen; (b) OA signal vs. step tuned CO laser wavelength for 1000 ppM NH₃ in nitrogen

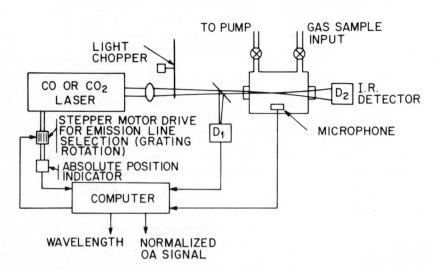

Figure 10. Steptuned CO_2/CO laser–OA pollution detection spectrometer

TABLE Ia

LASER LINES AND SENSITIVITIES				
Calibration Sample	NH_3 (1000 ppM)	NO_2 (100 ppM)	H_2O (10000 ppM)	HCN (1000 ppM)
CO Laser Line	$P_{16-15}(7)$	$P_{20-19}(14)$	$P_{24-23}(15)$	$P_{27-26}(12)$
$\nu(cm^{-1})$	1729.76	1605.32	1504.24	1442.15
$\lambda(\mu m)$	5.7812	6.2293	6.6479	6.9341
OA Signal	124	150	4.6	115

TABLE Ib

Interference Limited Sensitivity		
Component	Sensitivity Limit	Interference (Component)
HCN	0.5 ppM	5 % H_2O
NH_3	0.5 ppM	10000 ppM NO_2
NO_2	0.2 ppB	

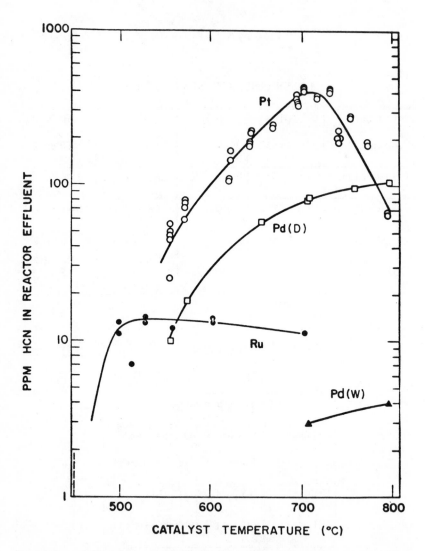

Figure 11.   HCN yield as a function of catalyst temperature for Ru, Pd, and Pt II catalysts. Pd(W) shows HCN yield for standard inlet gas mixture (5% CO, 0.5% H_2, 0.3% NO, balance He) with 3.5% added H_2O.

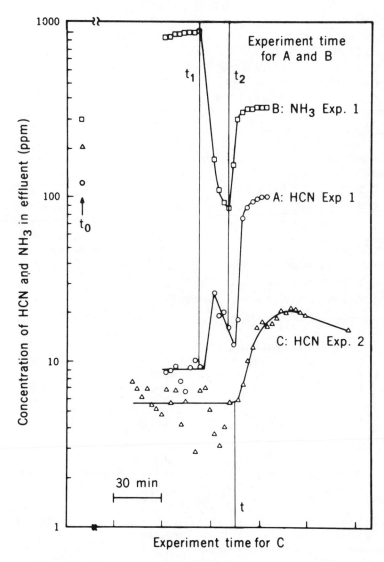

Figure 12.   Transient HCN yield over platinum catalysts. A and B represent one experiment (Exp. 1) in which a sulfur-poisoned catalyst was regenerated on admission of 1% O_2 into the inlet gas mixture between times t_1 and t_2. C shows the resistance of the catalyst to poisoning by SO_2 when oxygen is simultaneously present in the inlet gas mixture. At t only oxygen is removed from the inlet gas mixture. (See Ref. 16 for details.)

REFERENCES

1. See for example G. Herzberg, *Molecular Spectra and Molecular Structures,* Vols. I, II and III, (D. Van Nostrand).

2. L. B. Kreutzer, *J. Appl. Phys.* **42,** 2934 (1971); see also E. L. Kerr and J. A. Atwood, *Appl. Optics,* **7,** 915 (1968).

3. A. G. Bell, *Phil. Mag.* **11,** 510 (1881); J. Tyndall, *Proc. Roy. Soc.,* **31,** 307 (1881); W. C. Rontgen, *Phil. Mag.,* **11,** 308 (1881).

4. The use of acoustic resonances for increasing sound amplitude from an OA cell was first described by A. G. Bell, *Phil. Mag.,* **11,** 510 (1881). Recent experiments with resonant OA cells are described by C. F. Dewey, Jr., R. D. Kamm, and C. E. Hackett, *Appl. Phys. Lett.* **23,** 633 (1973).

5. C. K. N. Patel, *Phys. Rev. Lett.* **12,** 588 (1964); *Phys. Rev. Lett.* **13,** 617 (1964).

6. C. K. N. Patel, *Appl. Phys. Lett.,* **5,** 81 (1964); **6,** 12 (1965); **7,** 2736 (1965).

7. C. K. N. Patel and E. D. Shaw, *Phys. Rev. Lett.* **24,** 451 (1970); C. K. N. Patel in *Fundamental and Applied Laser Physics,* ed . M. Feld, A. Javan, and N. Kurnit, (Wiley, 1973) pp. 689-722; C. K. N. Patel in *Proc. Roy. Soc. (to be published);* see also C. R. Pidgeon and S. D. Smith, *Infrared Physics,* **17,** 515 (1977).

8. C. K. N. Patel and R. J. Kerl, *Appl. Phys. Lett.* **30,** 578 (1977).

9. L. B. Kreuzer and C. K. N. Patel, *Science,* **173,** 45 (1971).

10. H. L. Johnston, *Science,* **173,** 517 (1971); P. J. Crutzen, *J. Geophys. Res.* **76,** 7311 (1971).

11. C. K. N. Patel, E. A. Burkhardt, and C. A. Lambert, *Science,* **184,** 1173 (1974); E. A. Burkhardt, C. A. Lambert, and C. K. N. Patel, *Science,* **188,** 1111 (1975); C. K. N. Patel, *Optical and Quantum Electronics,* **8,** 145 (1976).

12. M. J. Molina and F. S. Rowland, *Nature,* **249,** 810 (1974).

13. L. B. Kreuzer, N. D. Kenyon, and C. K. N. Patel, *Science,* **177,** 347 (1972).

14. R. J. H. Voorhoeve, C. K. N. Patel, L. E. Trimble,and R. J. Kerl, *Science,* **190** 149 (1975).

15. I would like to thank Mr. R. H. Rabiner of Data General Corp. for the invaluble help and advice about the computer during the course of these experiments.

16. R. J. H. Voorhoeve, C. K. N. Patel, L. E. Trimble, and R. J. Kerl, *Science,* **200,** 761 (1978).

17. R. L. Bradow and F. D. Stump, SAE *(Soc. Automot. Eng.)* Tech. Paper 770367 (1977); U. S. Environmental Protection Agency, Office of Air and Waste Management, *Review of HCN Emission Data for 1977 Model Year Certification Vehicles* (Washington, D. C., 8 September 1976).

RECEIVED November 17, 1978.

Selective Ionization and Computer Techniques for the Mass Spectrometric Analysis of Air Pollutants

T. M. HARVEY, D. SCHUETZLE, and S. P. LEVINE

Research Staff, Ford Motor Company, Dearborn, MI 48121

Over the last several years there has been increasing concern about the potentially harmful effects that trace quantities of certain organic compounds might have on human beings and their environment. This concern spans many disciplines including identifying toxic substances and their metabolites, profiling manufacturing or vehicle emissions, and assessing air and water quality. Recent legislation has emphasized the need for evaluating the effect of chemicals on the environment (risk assessment). A significant part of risk assessment consists of identifying the nature, concentration, and sources of chemicals released into the environment as a result of man's activities. These tasks often require the identification and quantification of organic compounds below the parts per billion level in complex matrices.

A recent review (1a) emphasizes the complexity of air pollutant samples and describes some of the techniques which have been recently developed for their analysis. Among these are a number of alternate ionization techniques for mass spectrometry. Positive and negative chemical ionization mass spectrometry (CIMS) are analytical tools which can provide significant assistance in solving these problems. Their application in a number of air pollution studies are discussed in this chapter.

Air Pollution Studies

The chemical characterization of atmospheric pollutants is of great importance for determining their primary sources, elucidating chemical reactions in the atmosphere, determining potential risk to the environment and developing a reasonable control strategy.

Air pollution studies may encompass ambient and interior (work space) air environments and source emissions. Chemical compounds present in the atmosphere include gaseous materials, vapors, liquid particulates, solid particulates, and combinations of those phases. Air pollutants are classified as primary or secondary in nature, depending upon their sources. Primary air pollutants include natural (i.e., vegetation) and anthropogenic (i.e., stacks and vehicles) sources. Atmospheric reactions may generate secondary products from primary pollutants. For instance, the photochemical smog formed during atmospheric inversions in Los Angeles and other cities is the result of secondary air pollutant formation. It has been shown in a recent study (1b), that one reactive gaseous pollutant can form more than thirty different particulate reaction products.

0-8412-0480-2/79/47-094-195$05.25/0

The relationship of particle size to molecular composition is highly significant. (2). Particles below 1-3 μm (submicron size class) in diameter are considered respirable by animals, in that this material can reach the lower alveoli of the lung. Particle size information may be used to determine the sources of various atmospheric pollutants. Particulate matter formed from gaseous pollutants tends to be found in the less than the 1-3 μm particle diameter size range. Supermicron sized particles (particles greater than 1- 3 μm) typically originate from primary emission sources. (i.e., stack emissions, vehicles, soil).

Many air pollutants exist in both the particulate and gaseous phases simultaneously, depending upon their vapor pressure and polarity. Atmospheric particulate matter may be defined as dispersed solid or liquid matter in a gaseous medium ranging from clusters of a few molecules to particles of about 20 μm in radius. The gaseous pollutants are discrete molecules. Some gaseous pollutants may be absorbed in particulate matter, depending upon the structure of the particulate matter (i.e. graphitic carbon content) and the polarity of the gaseous molecules.

Most atmospheric pollutant samples are very complex in nature, containing hundreds of different molecular species and a large range of concentrations. Because of the complexity of most air pollution samples, gas chromatography alone is not suitable for elucidating their composition. Analysis by mass spectrometry in conjunction with gas chromatography or other preseparation techniques is often necessary for the definitive analysis of air pollution samples. Consequently the use of mass spectrometry as a tool for the analysis of air pollution samples has grown rapidly during the past five years. Many new innovative techniques have been developed during this period of time. However, there still exist many problems which require a solution. The successful analysis of air pollution samples depends upon the number of important steps schematically illustrated in Figure 1. This figure incorporates most of the major analytical techniques used to date for the mass spectrometric analysis of air pollutants. The major steps include sampling, pretreatment, and instrumental analysis. A number of recent reviews describe various aspects of the techniques (3-6) presented in this figure. The purpose of this paper is to describe new computer and ionization techniques which are currently being developed for the mass spectrometric analysis of air pollutants.

Traditionally, conventional electron impact techniques have been used to effect sample ionization. The often complex fragmentation pattern which results can provide invaluable information for elucidating the structure of unknown compounds. Vast computer libraries of electron impact spectra are available to assist the analyst with identifications. However, electron impact is an ionization technique that is somewhat limited in scope. By contrast, sample ionization by chemical ionization techniques offers the potential for selectivity, increased sensitivity, complimentary fragmentation, and confirmation or determination of molecular weight. (7). In this paper we will describe several examples in which the versatility of the chemical ionization technique has simplified or improved an analysis.

Chemical Ionization Principles

Several recent reviews can provide the reader with an historical perspective and information about additional applications of chemical ionization mass spectrometry. (7-10) Under normal conditions, electron impact mass spectrometry is carried out by bombarding sample molecules with 70 eV electrons in a mass spectrometer ion source

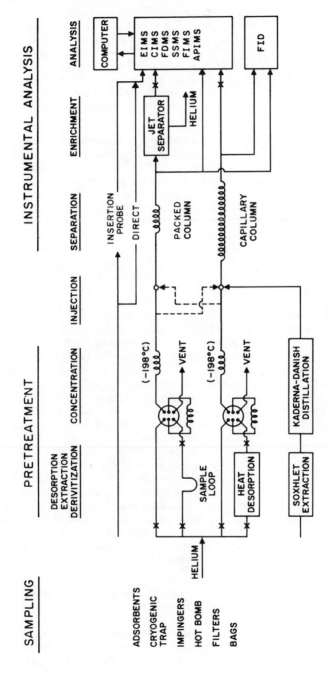

Figure 1. Summary for a variety of sampling, pretreatment and instrumental variations used in the GC and GC/MS analysis of air pollutants

operated at 10^{-7} to 10^{-6} torr. The interaction of the electron with the molecule usually produces radical cations which may contain upwards of 150-200 Kcal/mole excess internal energy. Through a variety of pathways, many with rather low activation energies, these molecular radical cations can fragment to smaller ions or neutrals while still resident in the ion source. These ions are removed from the source and mass analyzed using any one of a number of devices (magnetic sector, quadrupole, time of flight, ion cyclotron, etc.) to produce the familar mass spectrum. While the spectrum will likely contain a wealth of structural information, often it is difficult to assign the molecular weight of the compound. In addition, little or no selectivity of ionization can be achieved using electron impact analysis.

Analysis of organic molecules in the chemical ionization mode is accomplished in a slightly different manner. First, the apertures of the ion source must be reduced in size to decrease their gas conductance. A reagent gas can then be added to the ion source at pressures in the region of 1 torr. Such a gas load requires that the ion source region and the analyzer region be differentially pumped to maintain a long mean free path in the analyzer region. In addition, the pumping speed in the source housing is increased to provide effective pumping in the region of the ion exit slit and electron entrance hole sufficient to minimize ion/neutral scattering in the lens system. Once the reagent gas is in the ion source, it is bombarded with electrons to produce the common electron impact fragments. The increased ion source pressure results in considerable reduction in the mean free path of a primary radical cation within the source. Each ion or neutral in the source will undergo 20-1000 collisions (depending upon source design and gas conductance) before exiting to the high vacuum region. As shown in Figure 2, this results in the creation of a steady state plasma within the source, which for methane consist of CH_5^+, $C_2H_5^+$ and $C_3H_5^+$. These ions are unreactive to further collisions with neutral methane but can undergo ion-molecule reactions with compounds of interest which are added to the ion source. If the concentration of the compound being analyzed is allowed to rise higher than a few parts in 10^3, the potential for direct ionization by electron bombardment increases. However, if the sample pressure is maintained at an appropriate level, ionization of sample molecules will be effected by one of several ion-molecule reaction mechanisms (Figure 3). In the case of CH_5^+, ionization can occur via a proton transfer mechanism from CH_5^+ to a molecule having a higher gas phase proton affinity than methane (i.e., to a poorer Bronsted acid) to produce MW + 1. On the other hand, C_2H_5 can abstract a hydride to produce neutral ethane and a sample ion at MW − 1. Both $C_2H_5^+$ and $C_3H_5^+$ can undergo electrophilic addition to appropriate molecules to produce ions at MW + 29 or MW + 41. Chemical ionization spectra in contrast to electron impact spectra are usually characterized by a larger percentage of the total sample ion current being concentrated in an ion characteristic of the molecular weight. This is illustrated in Figure 4 comparing the electron impact and chemical ionization spectra of methyl stearate. Such behavior is in part a result of the lower exothermicity of the ionization process and the more limited number of decomposition pathways available to the even electron species.

One of the most powerful attributes of the chemical ionization technique is its flexibility. We are by no means limited to using methane as a reagent gas. Hydrogen, (11) propane, (12) ethane, (12) isobutane, (13,14) hexane, (15) ammonia, (16,17) methylamine, (16,17) dimethylamine, (17) nitric oxide, (18,19) argon, (20) H_2O, (20,21) D_2O, (22) nitrogen, (23) carbon monoxide, (23) methyl nitrite, (24) helium, (25) and numerous other gases have been successfully employed as chemical ionization reagent gases. In

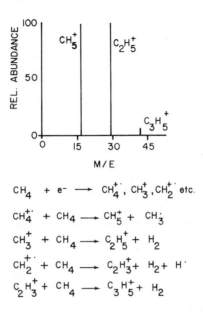

$$CH_4 + e^- \longrightarrow CH_4^{+\cdot}, CH_3^+, CH_2^{+\cdot} \text{ etc.}$$

$$CH_4^{+\cdot} + CH_4 \longrightarrow CH_5^+ + CH_3^{\cdot}$$

$$CH_3^+ + CH_4 \longrightarrow C_2H_5^+ + H_2$$

$$CH_2^{+\cdot} + CH_4 \longrightarrow C_2H_3^+ + H_2 + H^{\cdot}$$

$$C_2H_3^+ + CH_4 \longrightarrow C_3H_5^+ + H_2$$

Figure 2. Production of chemical ionization reagent ions with methane reagent gas

POSITIVE ION FORMATION

A) ELECTRON IMPACT

$$RXH + e^{\circ} \longrightarrow RXH^{+\cdot} + 2e^{\circ}$$

B) PROTON DONATION (BRONSTED ACID)

$$RXH + CH_5^+ \longrightarrow RXH_2^+ + CH_4$$

C) NEGATIVE ION ABSTRACTION

$$RXH + C_2H_5^+ \longrightarrow RX^+ + C_2H_6$$

D) ELECTROPHILIC ADDITION

$$RXH + C_2H_5^+ \longrightarrow \overset{+}{R}XH$$
$$\qquad\qquad\qquad\quad | $$
$$\qquad\qquad\qquad\; C_2H_5$$

E) CHARGE EXCHANGE

$$RXH + A^{+\cdot} \longrightarrow RXH^{+\cdot} + A$$

Figure 3. Mechanisms for positive ion formation in a chemical ionization source

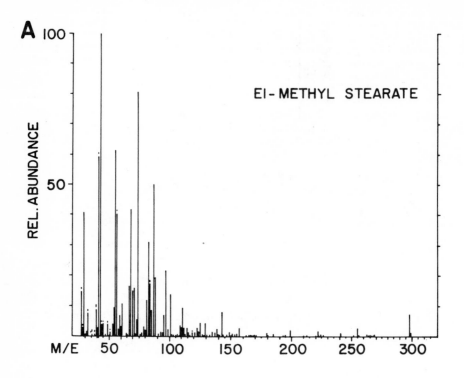

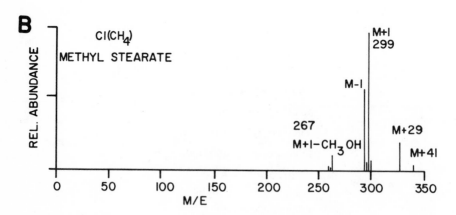

Figure 4. (a) Electron impact (70 eV) and (b) chemical ionization (CH_4) spectra
of methyl stearate

most cases each was chosen to fulfill a particular analytical need utilizing the unique ion-molecule reaction of the individual gas.

In recent years the realm of chemical ionization analysis has been extended to include negative ion chemical ionization. (25-28) Historically, analysis of negative ions produced under electron impact conditions has been largely ignored because of poor sensitivity. Negative ions can be produced in the chemical ionization source by a variety of mechanisms as shown in Figure 5. For many classes of compounds, analysis by negative ion CI in the electron capture mode can provide an increase in sensitivity of two orders of magnitude compared to positive ion EI or CI. Negative ions can also be produced by nucelophilic addition, positive ion abstraction or charge exchange mechanisms usually with sensitivities comparable to those found with positive ion chemical ionization but employing different ion-molecular chemistry. Unfortunately, the utility of various reagent gases for negative ion chemical ionization has not been explored as extensively as for positive ion analysis.

Instrumentation

Analyses were carried out on a Micromass MM-16 single focussing mass spectrometer. Electron impact spectra were acquired at a source pressure of 1×10^{-6} torr utilizing 70 eV electrons regulated to a trap current of 100 μamps. Source temperature was 200°C. Chemical ionization spectra were acquired at a source pressure of 0.4 to 0.8 torr with 150 eV electron emission regulated to a current of 1000 μamps. The source temperature was 100-160°C. Sample introduction was accomplished by using a direct insertion probe for filters, a directly coupled inlet for capillary gc columns and a glass jet separator for packed gc columns. The mass spectrometer was operated at 1200 resolution (10% valley) with scan speeds of approximately 1 second per decade. Control, data acquisition and data processing were provided by an Incos 2000 data system.

CI Studies of Sulfuric Acid and Sulfate Salts

During the combustion of fossil fuels, the sulfur present as organosulfur compounds may be oxidized to SO_2 and SO_3 (or H_2SO_3 and H_2SO_4 in the presence of water), depending upon such factors as combustion temperature, oxygen concentration and the presence of catalytic materials in the exhaust or venting systems. After being emitted into the atmosphere, SO_2 and SO_3 can be converted to many species including H_2SO_3, H_2SO_4, $(NH_4)_2SO_4$, $NH_4 HSO_4$, $(NH_4)_3H(SO_4)_2$ and NH_4HSO_3 in the presence of atmospheric water and ammonia. It is important to distinguish among these species for evaluating their environmental impact.

Several investigators have utilized thermal techniques for the separation of sulfate species collected on filter media with subsequent analysis by electron impact mass spectrometry, wet chemical analysis or sulfur flame photometry. In most instances the separation between sulfuric acid and its ammonium salts was incomplete or problems were encountered in recovering the species of interest from filters heavily laden with particulate (29-34).

Electron impact mass spectral analysis of sulfuric acid and sulfate salts (most notably ammonium sulfate) from filter samples has not been totally successful. Although the sensitivity of the technique was quite high (10^{-9} gms), both sulfuric acid and ammonium sulfate gave only SO_3^+, SO_2^+, SO^+ fragments when analyzed by electron impact mass spectrometry. Since their spectra were very similar, accurate quantitation would require that only one species be introduced into the source at a

NEGATIVE ION FORMATION

A) ELECTRON CAPTURE
$$x + e^\bullet \rightarrow x^-$$

B) NUCLEOPHILIC ADDITION
$$x^- + y \rightarrow xy^-$$

C) POSITIVE ION ABSTRACTION
$$x\text{-}y + z^- \rightarrow x^- + y\text{-}z$$

D) CHARGE EXCHANGE
$$x^- + y \rightarrow x + y^-$$

Figure 5.   *Mechanisms for negative ion formation in a chemical ionization source*

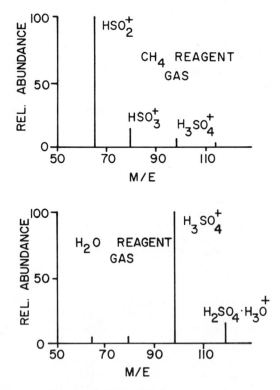

Figure 6.   *Chemical ionization spectra of sulfuric acid using methane and water reagent gases*

time. The filter samples were introduced into the instrument via a direct insertion probe which could subsequently be heated at various rates. Thermal separation of the species (attempted with a wide variety of temperature programming schemes) did not prove adequate for quantitative work.

It has been observed in this laboratory, as in many others, that compounds, which volatilize from the mass spectrometer direct insertion probe at a certain temperature under electron impact conditions, will often volatilize at a lower temperature (50°C or more) if the analysis is carried out under chemical ionization conditions with reagent gas sweeping around the probe tip and then into the source. For sulfuric acid deposited on a Fluoropore filter, a decrease in volatilization temperature of approximately 70°C has been observed for sulfuric acid when the analysis is carried out in the chemical ionization mode. The elution profile is sharper and better defined than that obtained under electron impact conditions.

In our positive ion chemical ionization experiments, our prime concern was the relative gas phase Bronstead acidities of H_2SO_4, SO_2 and SO_3 and whatever was chosen as a reagent gas. The ionizing reaction in these cases was proton transfer from a reagent ion to the neutral molecule. The proton affinities for several compounds and potential reagent gases have been determined by Munson et al [35] and are listed in Table 1. (Proton affinity is defined as the negative of the heat of reaction for acquisition of a bare proton by a neutral molecule. It is therefore, inversely related to gas phase Bronstead acidity). From this table it is clear that if methane is used as a reagent gas, H_2SO_4, SO_2 and SO_3 will all be ionized via proton transfer from the methane reagent ions. If water is used, however, only H_2SO_4 will be ionized. The water and methane chemical ionization spectra of H_2SO_4 are shown in Figure 6. Note the differing fragmentations reflecting the differences in the exothermicity of the proton transfer reactions between methane or water and sulfuric acid.

Table 1

Species	Proton Affinity (Kcal/mole)
CH_4	118 ± 3
SO_3	134 ± 5
SO_2	155 ± 3
H_2O	165 ± 3
H_2SO_4	168 ± 3

Aqueous solutions of ammonium sulfate and ammonium bisulfate were deposited on Fluoropore filters, placed in the direct insertion probe, and analyzed in the chemical ionization mode (H_2O reagent) gas. The samples were heated from 100°C to 330°C at 15°C/minute. No sample ions were observed under these anlaysis conditions, even when several micrograms of ammonium salts were analyzed. The thermal decomposition of ammonium salts of sulfate has been the subject of many studies. [29,30] Some pathways include sulfuric acid production at one stage of the decomposition while others suggest ammonia, SO_2 and SO_3 are the products. None of these accurately simulate the conditions (temperature, pressure, gas flow) present in our chemical ionization source. However, no sulfuric acid ions ($H_3SO_4^+$, etc.) were ob-served during CI (H_2O) analyses of ammonium salts of sulfate. This suggests that sulfuric acid is not produced from the decomposition of ammonium sulfate salts in the ion source.

Methane chemical ionization spectra of ammonium sulfate and ammonium bisulfate are similar to that of sulfuric acid. The elution profile, however, is quite different, as can be seen from Figure 7. If the salts were deposited on the filter in aerosol form, the shape of the elution profile was somewhat narrower and began at a slightly lower temperature compared to the profile obtained when the salt was deposited in aqueous solution from a syringe. In no case, however, did the elution profile of any of the salts tested overlap the profile for sulfuric acid.

For analysis of any filter sample heavily laden with particulate and/or hydrocarbon residues there is one further refinement necessary. Since the mass spectrometer is operated a 1200 dynamic resolution, ions occurring at m/e 65.039 ($C_5H_5^+$) and 64.970 (HSO_2^+); 81.070 ($C_6H_9^+$) and 80.965 (HSO_3^+); 99.117 ($C_7H_{15}^+$) and 98.975 ($H_3SO_4^+$) are easily resolved. These data are acquired with an accuracy of $\pm$ 25-30 ppm under our scanning conditions thus allowing the production of accurate mass chromatograms for the ions of interest. This is illustrated by an example in Figure 8. This analysis was carried out using methane as the reagent gas. We can see that the profiles for $H_3SO_4^+$ and $C_7H_{15}^+$ are quite different. Merely observing nominal mass chromatograms could have led to serious error in the determination of H_2SO_4.

Analyses of ambient air samples have thus far failed to detect the presence of sulfuric acid. However, considerable quantities of ammonium sulfate salts have been detected. One possible explanation is that sulfuric acid aerosol trapped on a filter is converted to ammonium salts by reaction with ammonia in the air pulled through the filter. A laboratory generated sulfuric acid aerosol collected on a Fluoropore filter was placed in a filter holder. Arbitrarily selected suburban and urban air was passed through the filter at a rate or 30 liters/minute for approximately one hour. In every case > 95% of the sulfuric acid was apparently converted to ammonium salts of sulfate. A strict material balance was not performed. A blank sample of laboratory generated sulfuric acid aerosols was transported to and from the field with proper precautions. Less than 5% conversion of the sulfuric acid to ammonium sulfate was observed for this sample.

Preliminary experiments suggest that negative ion chemical ionization may provide additional information in dealing with this problem. Negative ion chemical ionization spectra of sulfuric acid and ammonium sulfate obtained using isobutane as the reagent gas consist of SO_4^- (m/e 96) and SO_3^- (m/e 80) ions. In this configuration we are concerned not with the Bronstead acid capabilities of $C_4H_9^+$ but with the ability of isobutane to create a significant population of thermal or low energy electrons. These electrons can then be captured by sample molecules entering the source to produce molecular anions and their fragments. In the case of ammonium sulfate and sulfuric acid the negative ion chemical ionization spectra are quite similar. They are, however, distinct from the negative ion spectra of SO_2 and SO_3. These results suggest that a combination of positive and negative ion chemical ionization can provide a significant speciation of sulfur oxides and sulfates through differences in volatilization characteristics and mass spectral fragmentation patterns.

Analysis of Polycyclic Aromatic Hydrocarbons

Another area where negative ion chemical ionization has been applied in our laboratory is for the analysis of polycyclic aromatic hydrocarbons. Separation of these

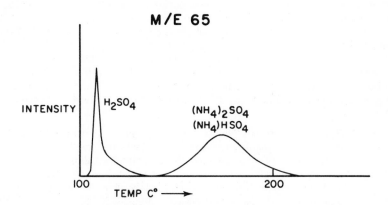

Figure 7.   *Volatilization profiles of sulfuric acid and ammonium salts of sulfate from solids probe using CI(CH₄) mass spectrometry*

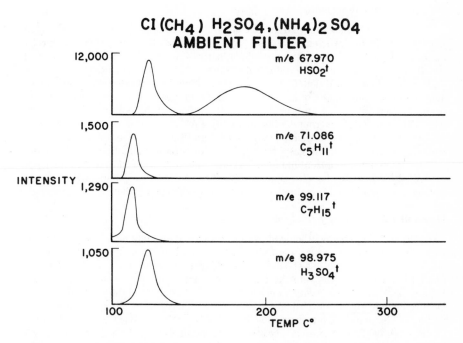

Figure 8.   *Accurate mass chromatograms—CI(CH₄) analysis (solids probe) of an ambient filter spiked with H_2SO_4 and ammonium sulfate salts*

compounds from the other less interesting but often major matrix components is a task which may complicate their analysis by gc or gc/ms. Many polycyclic aromatic hydrocarbons, particularly those with five or more rings or those containing heterocyclic atoms, have been found to exhibit high electron capture cross sections.[24] In some instances observed detection limits for selected PAH's using negative ion CI have been reported which are one to two orders of magnitude lower than those obtained by either positive ion CI or positive ion EI. Conversely, aliphatic hydrocarbons and many alkyl benzenes show very poor electron capture efficiencies. These results provide a means of selectively ionizing many PAH's in the presence of excess hydrocarbon residue.

As an illustration, consider Figures 9 and 10. Figure 9 illustrates the total ion current trace obtained from the positive ion chemical ionization (methane reagent gas) analysis of a sample of a residue obtained from distilling a gallon of gasoline (no additives). This sample contains many organic constituents normally found in atmospheric particulate matter and is often used in our laboratory for methods development purposes. The analysis was carried out on a 50 meter X 0.25 mm OV-101 glass capillary column programmed from 110° to 280° at 2°/minute with helium flow of 20 cm/second. The chromatogram is quite complex. If the analysis is carried out in the negative ion chemical ionization mode, the total ion current chromatogram in Figure 10 is obtained. While there is significant suppression of the aliphatics and alkyl benzenes, intensities for many polycyclic aromatic hydrocarbons are enhanced by an order of magnitude or more

Often time-consuming extractions, separations, column preparations, etc., are necessary to insure efficient analysis by gc/ms. Another technique involves survey analyses of particulate samples, particulate extracts, or adsorbent trap extracts using the direct insertion probe. The sample of interest is merely inserted into the ion source and heated at an appropriate temperature programmed rate. Continuous scanning or selected ion monitoring data may be acquired. From these analyses one can generate accurate mass chromatograms for the molecular anions of the polycyclics of interest. The high signal to background ratio for many PAH's in the negative ion CI mode makes a probe analysis of such a complex mixture feasible. As we see in Figure 11, a solid probe analysis of the gasoline residue affords us a qualitative picture of the polycyclics which are present in much less time than the gc/ms analysis. We have not evaluated the negative ion CI responses for all the polycyclic aromatic hydrocarbons nor have we attempted quantitative analysis from the solid probe.

Qualitative Class Survey Analysis

Negative ion chemical ionization can also provide qualitative class information from filter samples of atmospheric aerosols. A small piece of filter containing the aerosol sample is placed in a specially designed direct insertion probe and the components thermally desorbed into the ion source. Such techniques have been used by Schuetzle[36] et al, with positive ion electron impact analysis of the evolved species. Data reduction was accomplished by generating accurate mass chromatograms for a variety of characteristic low mass ions including Cl^+, NO_2^+, SO_2^+, $C_4H_9^+$, HCO^+, etc. From these data the presence or absence of certain classes of compounds could be inferred. The technique is rapid, sensitive and does not require any extensive sample workup. For several classes of compounds, negative ion electron capture analysis can provide improved sensitivity and specificity for these class analyses. Ions such as Cl^-, NO_2^-, CN^-, NCO^-, SO_4^-, SO_3^-, I^-, Br^-, and F^-, to mention a few, will be

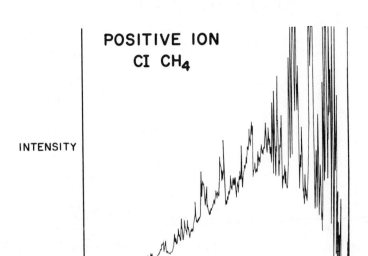

Figure 9. Total ion current plot from positive ion CI(CH₄) GC/MS analysis of gasoline residue (injection at right of plot)

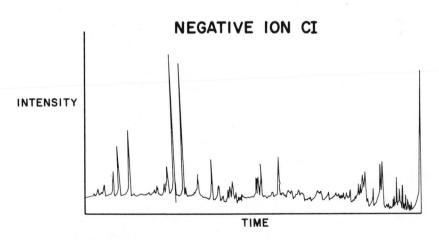

Figure 10. Total ion current plot from negative ion CI(CH₄) GC/MS analysis of gasoline residue (injection at right of plot)

PROBE NCI (MIXTURE)

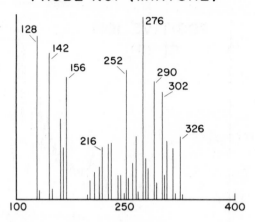

Figure 11. *Spectrum from solids probe negative ion CI(CH$_4$) analysis of gasoline residue*

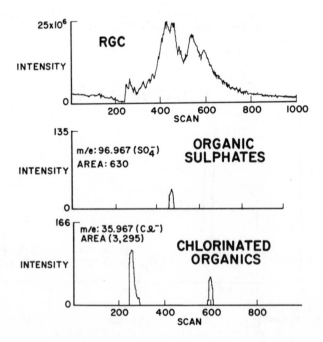

Figure 12. *Class searches from solids probe analysis of ambient filter sample-negative ion CI(CH$_4$) accurate mass fragmentograms*

produced in greater abundance from molecules containing no other strong electron capture moiety. Therefore class searches from accurate mass chromatograms can be performed with greater selectivity and greater sensitivity than can be accomplished with positive ion techniques. Figure 12 indicates the presence of chlorinated species as well as sulfate in an ambient aerosol. This preliminary screening has been invaluable in determining the course of subsequent analyses of ambient filter samples.

Analysis of Aldehydes in Complex Matrices

In many manufacturing processes there exists the potential for aldehyde formation. Often these aldehydes occur in low concentrations in the presence of much higher levels of aliphatics, olefinics and aromatic hydrocarbons. Gas chromatography or combined gc/ms methods are often ineffective in determining aldehydes in such a matrix. Several wet chemical techniques have been devised for estimating the total aldehyde concentration in these streams, but quantitation of the individual aldehydes has remained a difficult task.

A sampling and analysis scheme, utilizing the selective ionizing power of chemical ionization, has proven useful in approaching this problem. Air from the area being sampled is impinged into a pyridine solution of methoxylamine hydrochloride. Aldehydes present in the gas stream are subsequently converted to their o-methyl oxime derivatives.[37,38] Figure 13 displays the electron impact spectra of some of the aldehydes o-methyl oximes. Since these spectra differ from those of the underivatized aldehyde, combined gc/ms analysis of the impinger solution could provide qualitative identification of the aldehydes with less interference from aliphatics, olefinics, and aromatics. There are, however, still ions common to some of the oximes and interferences.

Quantitation and identification of the oximes is facilitated by using chemical rather than electron impact ionization. Figures 14 and 15 present the chemical ionization spectra of selected oximes using isobutane and ammonia, respectively, as reagent gases. In each case, there is an intense ion characteristic of the molecular weight. If isobutane is used as the reagent gas, however, there will be little discrimination against the ionization of potential hydrocarbon interferences. Figure 16 shows a total ion current trace obtained from the CI (ammonia) analysis of four oximes in pyridine spiked with a 50 fold excess of o-xylene, mestiylene and hexadecane. Ionization of the oximes is favored by a factor of 30 to 50 over the hydrocarbons. Careful control of ion source temperature and ammonia pressure can provide even higher (> 200:1) rejection of the alkyl benzenes and virtually eliminate ionization of aliphatics and olefinics. Here we are taking advantage of the fact that the gas phase proton affinities of the oximes are apparently higher than that of ammonia while those for hydrocarbons are lower, making their ionization much less favorable compared to the oximes.

Several other aspects of the method have been tested. Known concentrations of acrolein, heptaldehyde, and benezaldehyde were added to an impinger containing 10 ml of 2% methoxylamine hydrochloride and 100 liters of clean air were drawn through the impinger. Greater than 95% of the aldehydes were retained in the impinger as the oximes. This suggests that isotopically labeled aldehydes could be added to each impinger before collection to serve as an internal standard for quantitation of the individual aldehydes. The effect of a heavily contaminated sample gas stream on displacement of components from the impinger has not yet been thoroughly investigated. Preliminary experiments in selected emission gas streams indicated a detection

ELECTRON IMPACT

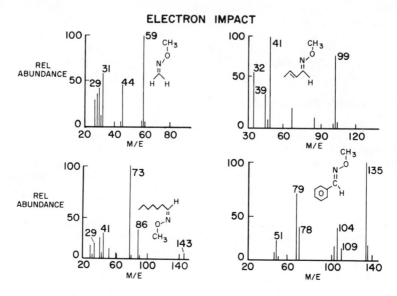

Figure 13.   Electron impact spectra (70 eV) of aldehyde o-methyl oximes

CI-ISOBUTANE

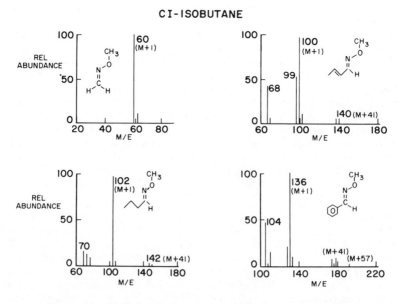

Figure 14.   CI (isobutane) spectra of aldehyde o-methyl oximes

CI- AMMONIA

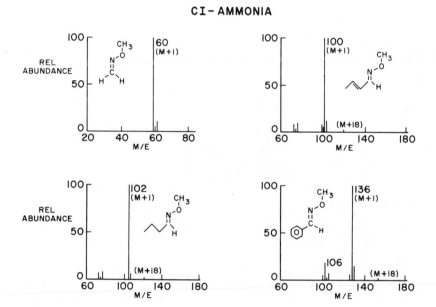

Figure 15. CI(NH₃) spectra of aldehyde o-methyl oximes

RGC CI(NH₃) EXCESS HYDROCARBONS

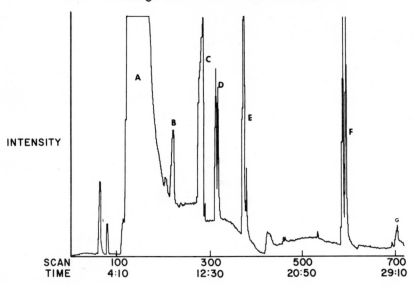

Figure 16.   Total ion current plot for the GC/MS analysis of selected aldehyde o-methyl oximes in the presence of fiftyfold excess hydrocarbons.   (Methoxime doublet peaks are from chromatographic resolution of syn and anti conformers.)

(A)   *pyridine (solvent)*
(B)   *xylene*
(C)   *mesitylene*
(D)   *heptaldehyde o-methyl oxime*
(E)   *benzaldehyde o-methyl oxime*
(F)   *cinnamaldehyde o-methyl oxime*
(G)   *n-hexadecane*

limit of 0.1 ppm can be easily achieved using a 10 ml impinger containing an internal standard. Insufficient data is currently available to accurately assess the precision of the method. The effect of water in the air being sampled has also been evaluated. Series of impingers were set up with water in impinger one and a solution of aldehydes and methoxylamine hydrochloride in pyridine in impinger two. One hundred liters of air were passed through the tandem impingers. The pyridine impinger was then analyzed by gc/ms along with an identical impinger which sampled only dry air. Water seemed to have no adverse effect on the o-methyl oximes of the aldehydes over this analysis time. Qualitatively, the impinger efficiency appears quite good, but accurate determinations are proceeding using permeation tube to generate standard vapor streams.

A drawback to this technique is the presence of the pyridine solvent. The air sampling pump must be protected from fouling with pyridine by the use of an impinger containing dilute sulfuric acid solution interposed between the sampling impinger and the pump. Secondly, the pyridine is not the ideal solvent for gc/ms analysis if aldehydes from C_3 to C_6 are to be analyzed. Tests will be performed in order to either eliminate the pyridine from the sampling solution or from the solution of oximes after sampling is complete. This should result in cleaner chromatograms both for gc/ms analysis and for gc separation coupled with the use of a nitrogen thermionic detector.

Summary

Chemical ionization mass spectrometry has been found to be an effective tool in our laboratory for the analysis of a variety of air pollution samples. The ability to control the ionization process by proper choice of reagent gas is valuable in a number of areas. These include selective ionization of certain interesting compounds in a complex matrix, increased reliability in molecular weight determinations, and less complex spectra, which may lead to better quantification. In addition, structure elucidation of unknown compounds may be facilitated because of CI fragmentations. The scope of the method can be expanded by using various dervitizing reagents during sample collection or workup. Judicious selection of a derivative can provide enhanced selectivity in the CI analysis of air pollution samples. The high sensitivity of mass spectrometry will continue to make it the method of choice for a variety of pollution studies. Proper use of CI techniques will provide the analyst considerable flexibility in combating a broad spectrum of problems.

REFERENCES

1a. Schuetzle, D., Air Pollutants, Chap 32B in Biochem. Appl. of Mass ,Spectroi, G. Waller, ed., in press.

1b. Schuetzle, D., and Rasmussen, R. A., J. Air Poll. Control Assoc., (1978), **28,** 236.

2. Katz, M., "Methods of Air Sampling and Analysis," American Public Health Assoc., Washington, D.C. (1977).

3. Alford, A., Biomedical Mass Spectrometry (1975), **2,** 229.

4. Burlingame, A. L., Shackleton, C. H. L., Cedric, H. L., Howe, I., and Chuzbov, O. S., Anal. Chem., Annual Reviews (1978) **50,** 371 R.

5. Alford, A., Biomedical Mass Spectrometry (1977) **4** (1).

6. Burlingame, A. L., Kimble, B. J., Derrick, P. J., Anal. Chem., Annual Reviews (1976) **48,** 368 R.

7. Field, F. H., Accounts Chem. Res., (1968) **1,** 42.

8. Field, F. H., in "Ion Molecule Reactions," Vol. 1, Franklin, J. L., ed., Plenum Press, New York, (1972) 261.

9. Munson, B., Anal. Chem. (1977) **49,** 772A.

10. Wilson, J. M., "Mass Spectrometry," Vol. 4, Johnstone, R. A. W., Sr. Reporter, Chemical Society, London (1977), 111.

11. Gillia, R. G., Lacey, M. J., Shannon, J. S., Org. Mass Spectrom. (1974) **9,**359.

12. Munson, M. S. B., Field, F. J., J. Amer. Chem. Soc., (1965) **87,** 3294.

13. Field, F. H., J. Amer. Chem. Soc. (1969) **91,** 2827.

14. Field, F. H., J. Amer. Chem. Soc. (1969) **91,** 6334.

15. Field, F. H., Yu, T. Y., Org. Mass Spectrom. (1974) **8,** 267.

16. Munson, M. S. B., J. Phys. Chem. (1966) **70,** 2034.

17. Hunt, D. F., McEwen, C. N., Upham, R. A., Tetrahedron Lett. (1971) 4539.

18. Hunt, D. F., Ryan, J. F., Chem. Commun. (1972), 620.

19. Jelus, B. L., Munson, B., Fenselau, C., Biomed. Mass Spectrom. (1974), **1,** 96.

20. Hunt, D. F., Ryan, J. F., Anal. Chem. (1972) **44,** 1306.

21. Buttrill, S. E., Upham, R. A., Price, P., Martinsen, D. P., Swofford, H. S., Anal. Chem. (1975) **47,** 190.

22. Hunt, D. F., McEwen, C. N., Upham, R. A., Anal. Chem. (1972), **44,** 1292.

23. Einolf, N., Munson, B., Int. J. Mass Spectrom. Ion Phys. (1972), **9,** 141.

24. Hunt, D. F., Stafford, G. S., Crow, F. W., Russell, J. W., Anal. Chem. (1976) **48,** 2098.

25. Bowie, J. H., Williams, B. D., "MTP International Review of Science, Physical Chemistry, Series Two," Vol. 5, A. Maccoll, Ed., Butterworth, London (1975) 89.

26. Dillard, J. G., Chem. Rev., (1973), **73,** 589.

27. Von Ardenne, M., Steinfelder, K., Tummler, R., "Electromenanlagerungs-Massenspektrographie Organischer Substanze," Springer-Verlag, New York (1971).

28. Jennings, K. R., in "Mass Spectrometry," Vol. 4, Johnstone, R. A., Sr. Reporter, Specialist Periodical Reports, Chemical Society, London (1971), 203.

29. Dubois, L., Baker, C. J., Teichman, T., Zdrojewski, A., and Monkman, J. L., Mickrockim. Acta (Wien): (1979) 269.

30. Maddalone, R. F., Shendrikar, A. D., and West, P. W., Mickrochimica Acta (Wien): (1974) 391.

31. Maddalone, R. F., McClure, G. M., and West, P. W., Anal. Chem., (1975), **47,** 316.

32. Mudgett, P. S., Richards, L. W., and Roehrig, J. R., "A New Technique to Measure Sulfuric Acid in the Atmosphere," in Analytical Methods Applied to Air Pollution Measurement, R. Stevens and W. Herget, Eds., Ann Arbor Science Pub., Inc. Ann Arbor, MI 1974.

33. Leahy, D., Siegel, R., Koltz, P., and Newman, L., Atmos. Environ. (1975), **9,** 219.

34. Huntzecker, J. J., Isabelle, L. M., and Watson, J. G., presented at the 69th Annual Meeting of the Air Pollution Control Association, Portland, Oregon, June, 1976, and references, therein.

35. Munson, M. S. B., private communication.

36. Schuetzle, D., Ph.D. Dissertation, University of Washington, Seattle, Washington (1972).

37. Gordiner, W. L., and Horning, E. C., Biochem. Biophys. Acta. (1966), **115,** 425.

38. Horning, M. G., Moss, A. M., and Horning, E. C., Anal. Biochem. (1968), **22,** 284.

RECEIVED November 17, 1978.

Sampling and Analysis for Semivolatile Brominated Organics in Ambient Air

RUTH A. ZWEIDINGER, STEPHEN D. COOPER, MITCHELL D. ERICKSON, LARRY C. MICHAEL, and EDO D. PELLIZZARI

Research Triangle Institute, Post Office Box 12194, Research Triangle Park, NC 27709

Introduction

Increasingly strict guidelines regarding the flammability of textiles, particularly synthetic materials used in infant garments, have initiated the development and use of more effective flame retardant chemicals. Among the flame retardants which have been widely used is tris(2,3-dibromopropyl)phosphate (TRIS). Although currently banned for domestic use, TRIS has been widely used in the past as a textile flame retardant and is still manufactured for export. The compound is extremely toxic to fish and has been demonstrated to have anticholinesterase activity [1]. Furthermore, recent tests have implicated it as a carcinogen [2] and mutagen (Chapter 1, this volume). Due to this imminent health hazard and its extensive historical usage, assessment of the introduction of TRIS and other brominated organics into the environment is of paramount importance.

The volatility of TRIS is very low, on the order of 4.8×10^{-3} Torr at $65°C$. This low volatility and the thermal instability of alkyl phosphates in general, particularly halogenated alkyl phosphates, presents a number of unique problems in the development of sampling and analytical methodology. The method development for TRIS was initiated with the evaluation of several detection methods. Taking into account its low volatility and relative thermal instability, sampling and analysis protocols were fomulated for the determination of TRIS. These protocols were extended to decabromobiphenyl ether (Decabrom), 2,2-bis(dibromo-4-hydroxy phenol) propane (Tetrabrom) and 1,2-bis(2,4,6-tribromophenoxy)-ethane (Firemaster 680®). Field samples were collected to validate the aforementioned protocols and evaluate the atmospheric levels of brominated organic aerosols in the vicinity of organo-bromine synthesis facilities.

Analytical Method Development for TRIS. The detection of brominated compounds of very low volatility such as TRIS posed special analytical problems. Since TRIS has no recognizable chromophore, the detection systems which are commonly used with high performance liquid chromatography (hplc), such as refractive index or short wavelength (<220 nm) uv detectors, are too non-specific to be of much practical use for the analysis of environmental samples. Furthermore, the sensitivities available with these detection methods are generally inadequate.

The inherent insensitivity of these methods prompted an evaluation of gas-liquid chromatography with electron capture detection for the analysis of TRIS. Due to the

0-8412-0480-2/79/47-094-217$05.00/0

low volatility, short columns (42 cm x 0.2 cm i.d.) were used. Two column packings were compared and found to produce similar resolution efficiency: 1.) 3% SE-30 on Chromosorb W(HP) and 2.) 3% OV-17 with benzyltriphenylphosphonium chloride on Chromosorb W(HP). Chromatography of TRIS at 204°C and 20 ml N_2/min on 3% SE-30 gave a retention time of 6 min (Figure 1). Decomposition of TRIS was observed with both column packings, however, a calibration curve (Figure 2) indicates that decomposition did not limit the analysis until injections of less than 0.5 ng were made. With this limit of detection, a minimum of ca. 20 ng/m^3 of TRIS could be quantitated. The glc/ecd peak at approximately 6 min was confirmed to be TRIS by gas-liquid chromatography/ mass spectrometry/computer (glc/ms/comp) and direct probe mass spectrometric analyses of pure compound using a Finnigan 3300 quadrupole gas chromatrograph/ mass spectrometer interfaced with a PDP-12 computer. The Finnigan 3300 has a mass range of 1000, with unit mass resolution over the entire range, and is capable of operating in the selected ion monitor (SIM) mode. The basic hardware of the PDP/12 consists of an 8K central processor fitted with a teletype, random access disc, CRT display and electrostatic printer/plotter. All data processing operations are carried out interactively by means of programs stored on the computer. The mass spectrum from m/e 430 to 650 is shown in Figure 3), where the characteristic bromine isotope clusters are apparent. The direct probe mass spectrum, shown in Figure 4, is nearly identical to that obtained by glc/ms and confirms that this compound is TRIS. Other peaks in the chromatogram were determined to be products of TRIS decomposition.

Glc/ms/comp Analysis of Brominated Organics. A Finnigan 3300 gc/ms with a PDP-12 computer was used for sample analysis. In order to obtain relatively narrow peaks of late eluting compounds (e.g. Decabrom) and avoid decomposition of TRIS, a short column was utilized. A 45 cm x 0.2 cm i.d. glass column packed with 2% OV-101 on 100/120 mesh Gas Chrom Q was temperature programmed from 220 to 300°C at 12°/min. OV-101 was substituted for SE-30 because of its stability at higher temperatures. The carrier gas (N_2) flow was 30 ml/min. Table I lists the compounds, their retention times and the m/z ions monitored. During sample runs the first and, where possible, second "ions" were monitored. Peaks having the correct retention times and ratio of intensities, where more than one mass was monitored, were confirmed by further analyses. With sufficient material, the sample was analyzed in the full scan mode and mass spectra obtained to confirm the identity of suspected brominated compounds. Without sufficient material, additional SIM analyses were performed and the intensity ratios were compared with those obtained for authentic compounds.

Compounds were quantified by comparing the computer calculated area for the brominated compound with the integrated response for a known amount of octachloronaphthalene. Differences in ionization cross-section, which affect the sensitivity of the mass spectrometer to a given compound, were compensated for by determining the relative molar response (RMR) of authentic compounds to octachloronaphthalene.

The calculation of the RMR factor allows estimation of levels of sample components without establishing a calibration curve. The RMR is calculated as the integrated peak area of a known amount of the compound, $Å_{unk}$ with respect to the integrated peak area of a known amount of standard, $Å_{std}$ (in this case octachloronaphthalene), according to the equation:

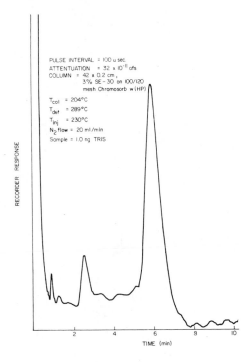

PULSE INTERVAL = 100 u sec.
ATTENTUATION = 32 x 10⁻¹¹ ofs
COLUMN = 42 x 0.2 cm,
3% SE-30 on 100/120
mesh Chromosorb w(HP)

T_{col} = 204°C
T_{det} = 289°C
T_{inj} = 230°C
N_2 flow = 20 ml/min
Sample = 1.0 ng TRIS

RECORDER RESPONSE

TIME (min)

Figure 1.   Gas chromatogram (^{63}Ni electron capture detection) of tris (2,3-dibromopropyl) phosphate standard

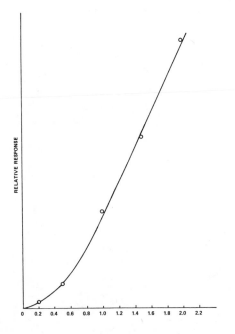

RELATIVE RESPONSE

Figure 2.   Calibration curve for the glc-ecd analysis of tris (2,3-dibromopropyl)-phosphate on 3% SE-30 on Chromosorb W(HP) 100/120 mesh

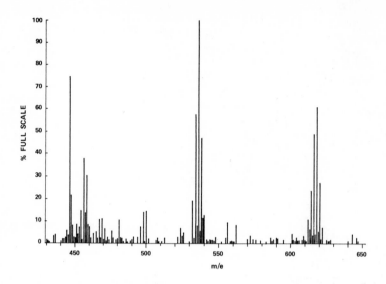

Figure 3.   Mass spectrum from GC/MS/COMP analysis of TRIS

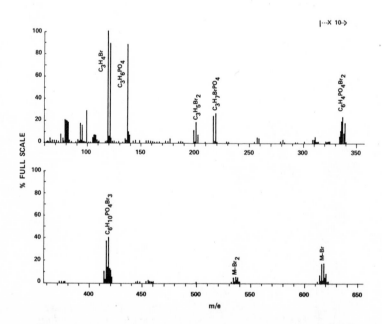

Figure 4.   Mass spectrum from direct probe introduction of TRIS

Table I. MID Ions Selected for Each Semi-Volatile Brominated Compound

Compound	Retention Time (min)	First[a] (m/e)	Second (m/e)	Third (m/e)
			Ions Monitored	
Octachloronaphthalene (External standard)	1.4	404		
Pentabromophenol	0.7	488	490	
Tetrabrom	2.2	529	531	
TRIS	2.1	417	419	
Firemaster 680	3.5	357	688	690
Decabrom	7.1	800	802	804

[a]For screening purposes, the ions in the first column plus three from the second were run.

$$RMR_{unknown/standard} = \frac{A_{unk}/moles_{unk}}{A_{std}/moles_{std}} = \frac{A_{unk}/g_{unk}/GMW_{unk}}{A_{std}/g_{std}/GMW_{std}} \quad \text{(Equation 1)}$$

From this calculated value, the concentration of an identified compound in a sample is calculated by rearranging Equation 1 to give

$$g_{unk} = \frac{A_{unk}.GMW_{unk}.g_{std}}{A_{std}.GMW_{std}.RMR_{unk/std}} \quad \text{(Equation 2)}$$

The use of RMRs for quantitation by glc/ms/comp has been used successfully in repeated applications to similar research problems (3-8). Typical RMRs listed in Table II are mean values of three injections of each of three replicate standard mixtures. They must be determined for each instrument and day-to-day variations are sometimes large enough to require daily recalibration. This is particularly true of compounds (e.g., Decabrom) for which the RMRs were calculated from peak areas generated by monitoring high mass ions where the instrument is less stable.

Thin-Layer Chromatography. Thin-layer chromatography was used for many of the compounds which are of interest for either qualitative or quantative analysis. These techniques were used primarily as screening methods to supplement other analytical methods. For some brominated compounds such as TRIS, the detection system reported by Hahn (9) was applicable. This method is based upon the reaction of fluorescein with elemental bromine released through peroxide oxidation. An evaluation in our laboratory of this detection system indicated that semi-quantitative results

Table II. Typical RMR[a] Values for Semi-Volatile Brominated Organics

Compound	Ion m/e	RMR[b,d]	S.D.	Ion m/e	RMR	S.D.
Pentabromophenol	488	0.142	0.039	490	0.141	0.036
Tetrabrom	529	0.410	0.066	531	0.279	0.048
Firemaster 680	357	0.752	0.114	688	0.0384	0.0126[c]
Decabrom	800	0.244	0.118	802	0.0494	0.0110[c]

[a]Relative molar response relative to the OCN 404 m/e ion.
[b]Average of 7 determinations.
[c]Average of 2 determinations.
[d]Value for TRIS not given due to poor sensitivity for these ions.

may be obtained using the technique of fluorescence quench thin-layer scanning. The basis for this detection method is the conversion of fluorescein impregnated in the thin-layer to the brominated analog, eosin. The resulting loss of fluorescence was then detected on a Schoeffel SD 3000 spectrodensitometer (sd) operated in the reflectance mode (tlc/sd). An example of a chromatogram obtained in this manner is shown in Figure 5. This detection system is highly specific for brominated compounds; however, the limit of detection is approximately 0.5 μg/spot. Some improvement in sensitivity may be obtained by optimizing chromatographic conditions. Basic elements of the reaction are acidic medium (glacial acetic acid), an oxidant [either H_2O_2 or chloramine-T (N-chloro-p-toluenesulfonamide)-sodium] and fluorescein.Thin-layer parameters for several semi-volatile brominated organics used throughout the program are given in Table III.

Table III
Thin-Layer Solvent Systems for Selected Brominated Organics

Compound	Solvent	R_f
Decabrom	Hexane/Toluene (95:05)	0.7
2,4,6-Tribromophenol	Toluene	0.65
Pentabromophenol	Toluene	0.5
Tetrabrom	Toluene	0.4
4-Bromophenol	Toluene	0.2
Firemaster 680	Hexane/Toluene (60:40)	0.6

Precision and Accuracy. For the glc/ms/comp analysis, the analytical precision may be estimated from the standard deviation of RMR factors generated from standards. Twenty-one separate determinations of five selected compounds over a nine-day period yielded a relative standard deviation of 24%. Actual results may be somewhat better since daily recalibration compensated for variations in instrument response.

Estimation of accuracy is based upon knowledge of analytical recoveries and instrumental accuracy. Standardization of the glc/ms/comp system was achieved by injecting three sets of standard mixtures at varying relative concentrations to the

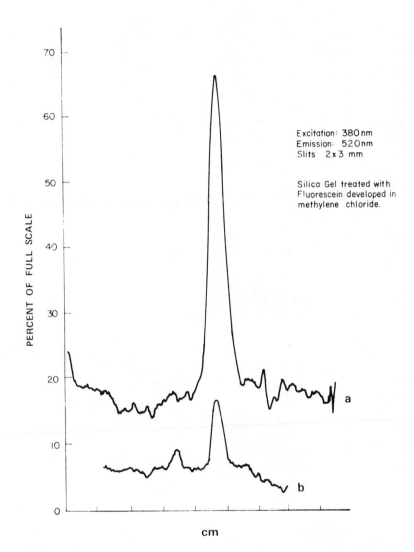

Figure 5.   *Thin-layer chromatogram of* tris *(2,3-dibromopropyl)phosphate on fluorescein impregnated silica gel; solvent: methylene chloride.   Scan in fluorescence quench mode.   (a) 3.7 μg TRIS; (b) 0.44 μg TRIS.*

internal standard. Variability between these was no greater than day-to-day instrument sensitivity fluctuation. Recoveries of TRIS and Decabrom by solvent extraction of glass fiber filters are shown in Table IV.

Table V lists the detection limits for several semi-volatile brominated organics by glc/ms/comp, glc/ecd and tlc/sd. By far the greatest sensitivity to TRIS is attainable with glc/ecd. However, the versatility and identification capabilities of glc/ms/comp make it the method of choice for sample analysis. Consequently, the majority of the samples were analyzed by glc/ms/comp.

Sampling

Development of Air Sampling Methodology. Several collection systems for isolating TRIS from ambient air were investigated.

Table IV. Recovery of Brominated Organics From Glass Fiber Filters After Air Sampling

Compound	Amount Added[a]		% Recovered
	(ng/cm^2)	(ng/m^3)	
Tris	104	27	87[b]
	1110	286	91[b]
Decabromobiphenyl ether	0	0	ND
	96	24	145[c]
	512	130	107 ± 14[c]

[a]Compound applied directly to filter followed by sampling 2038 m^3 of air.
[b]Analyzed by glc/ecd.
[c]Analyzed by tlc.

Table V. Detection Limits for Selected Semivolatile Halogenated Organics

Compound	Detection Limit (ng/m^3)		
	GC/MS/COMP[a]	GC/ECD	TLC/SD
TRIS[d]	614	4	250
Decabrom[d]	20-100[b]	n.d.[c]	n.d.
Tetrabrom	10	n.d.	n.d.
Firemaster 680	10	n.d.	n.d.
Pentabromophenol	25	n.d.	n.d.

[a]Multiple ion detection
[b]Variable due to daily fluctuations
[c]Not determined
[d]Detection limit of TRIS and Decalrom is high due to decompositional losses in GC/MS transfer line.

The apparatus shown in Figure 6, used for determining recoveries of TRIS on various sorbents, was comprised of three basic sections. The glass beads in Section A were spiked with TRIS and heated to 100°C while maintaining a flow rate of 3.7 l/min air through the beads with a Nutech Model 221-1A air sampling pump (Nutech Corp.,

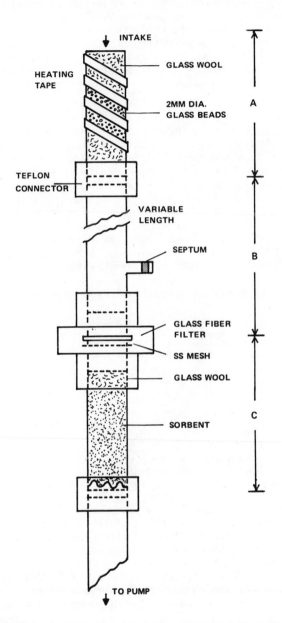

Figure 6.   Sample generation and collection apparatus

Durham, NC). Under these conditions the temperature of the air stream midway through Section B was 50°C and had dropped to 30°C upon reaching the filter (Section C). Aliquots of the air stream were withdrawn through the septum in Section B and analyzed by glc/ecd in order to determine the air stream concentration. After termination of air flow, the filter, glass wool plug and sorbent, if any, were extracted with acetone and the extracts analyzed by glc/ecd. The apparatus from the injection port to the filters was rinsed with acetone and analyzed to determine the TRIS which was surface absorbed.

The polystyrene resin, XAD-2, was evaluated as a sorbent for TRIS. An experiment in which the XAD-2 was located beyond a glass fiber filter in the apparatus described above revealed the TRIS is collected primarily on the filter (Table VI). A subsequent experiment with two glass fiber filters in series but without XAD-2 revealed similar results. The low recoveries may reflect the low volatility of TRIS, which may have caused it to condense on interior glass surfaces of the apparatus. In addition, the presence of other peaks in the thin-layer chromatographic analysis of extracts of the glass fiber filters suggested that decomposition may have occurred during the experiment. Cellulose filters (2.5 cm diameter, Whatman No. 1), which tend to be less reactive than glass fiber filters, were tested for collection efficiency. The results, shown in Table VI show not only higher recovery, but thin-layer chromatography of the cellulose filter extract indicated no discernable TRIS decomposition.

Table VI. Evaluation of Sampling Systems for Airborne TRIS

Media of Collection	Volume Collected (liters)	Average Flow Rate (l/min)	Recovery (μg) 1st Filter	Recovery (μg) 2nd Filter	Recovery (μg) Surface[e] Absorbed	Recovery (μg) Total	Amount of TRIS Predicted from Direct Analysis	Apparent Recovery of TRIS (%)
GFF[a]/XAD-2	363	3.60	84	<2[b]	1.2[c]	85	290	29
GFF	215	3.59	58	0.8	9.0	68	185	37
XAD-2	224	3.75	117[b]	—	11.3	128	454	28
Cellulose Filters	680	3.53	2.8	<0.1	7.1	9.9	68	14
Cellulose Filters	201	3.34	197	<0.14	18.5	215	434	50

[a]GFF = glass fiber filter
[b]XAD sorbent, not a second filter
[c]Glass wool plug. Collecting section was not evaluated
[d]XAD-2 sorbent without filter or glass wool packing
[e] Due to loss in apparatus

Preliminary air sampling at Chemical Company A in April, 1977, was conducted using two adjacent high-volume samplers (General Metals Works, Model GMWL-2000), one fitted with a cellulose filter (8 cm x 10 cm) in series with a glass fiber filter and the other with a glass fiber filter. For reasons not considered previously, the glass fiber filter was superior to the cellulose filter. First, the flow rate through the cellulose filters was 60% of that through the glass fiber filter and second, visual inspection of the filters revealed considerable breakthrough of particulate through the cellulose filter. On this basis, glass fiber filters were used in all subsequent sampling.

Field Sampling and Analysis. Air-borne particulate was collected on glass fiber filters using high-volume samplers. The exposed glass fiber filter was extracted with acetone and analyzed by either glc/ecd, glc/ms/comp or thin-layer chromatography. The sampling and analysis protocols developed specifically for TRIS were extended to

Decabrom, Tetrabrom, and Firemaster 680. All of these compounds are amenable to analysis by selected ion monitor glc/ms/comp. However, substantially lower detection limits for TRIS were attained using glc/ecd. Thin-layer chromatography was used for the analysis of Decabrom in some air samples collected during a preliminary study.

Blank and spiked samples were created by pipeting standard amounts of TRIS and Decabrom onto 10 cm^2 areas (2 x 5 cm) of glass fiber filter leaving blank (2 x 5 cm) blocks between each spiked area. The filter was then used to collect a 2038 m^3 air sample in the Research Triangle Park, NC. This filter was shipped to the field and returned for analysis. The results of analysis of areas of the filter are shown in Table IV. These results indicate the overall recoveries for sampling, shipping and storage as well as extraction recovery from this filter.

Air samples were collected on plant property of two industrial organo-bromine synthesis facilities, Companies A and B. After identification of suitable power sources the high-volume sampler was located so as to take advantage of meteorology, topography and proximity to synthetic facilities. The air sampling protocols are presented in Tables VII and VIII. A glass fiber filter (20 x 25 cm), pre-inspected for holes, was placed in the holder. The sampler was turned on, the time recorded and the air flow adjusted to between 1100-1200 l/min. After 24 hours, air flow was stopped and the time and flow rate recorded. The filter was removed from the holder, folded (exposed sides inward), wrapped in aluminum foil and gently rolled for insertion into a mailing tube. Upon arrival back in the laboratory the samples were logged into a notebook and stored at room temperature.

Table VII. High-Volume Air Sampling at Company A

Date	Volume Collected (m³)	T(°C)	% RH	Wind Direction Speed (kmph)
			Meteorological Conditions	
12/19/76	1364[a] 2404[b]			
12/21/76	1306[a] 2324[b]			
12/22/76	1463[a] 2604[b]			
7/22/77	1573[b]	18-38	52-88	NW @ 5-8 (4 hrs) S @ 0-3 (18 hrs)
7/23/77	1587[b]	21-40	50-92	N,NW @ 0-8 (13 hrs) S @ 2-8 (12 hrs)
7/24/77	1617[b]	21-38	45-92	S @ 3-11 (15 hrs) N,NE @ 4-5 (5 hrs) Variable @ 5 (5 hrs)
7/25/77	1438[b]	20-37	56-90	S,SE @ 2-12 (~16 hrs) NE @ 4-10 (~5 hrs)

[a]Collected on cellulose filter
[b]Collected on glass fiber filter

Table VIII. High-Volume Air Sampling at Company B

Date	Volume Collected (m³)	T(°C)	% RH	Wind Direction Speed (kmph)
			Meteorological Conditions	
12/13/76	1560[a] 1936[b]			
12/16/76	1577[a] 2584[b]			
12/17/76	1564[a] 2734[b]			
7/18/77	1949[b]	20-35		E,SE @ calm
7/19/77	2051[b]	20-32	90	N,NE @ calm NE,W @ 3-7
7/20/77	1928[b]			
7/21/77	1475[b]			

[a]Collected on cellulose filter
[b]Collected on glass fiber filter

Prior to analysis, the exposed glass fiber filters were carefully cut into 4 x 10 cm segments taken from the central area of the filter. The segments were further cut into narrow strips (~5 mm wide x ~4 cm long) which were placed collectively in a 20 ml vial containing 10 ml acetone. The vial was placed in an ultrasonic bath for 30 min or shaken on a reciprocal shaker at ~120 cpm for 2 hrs. An aliquot of the supernatant was withdrawn for analysis.

Results and Discussion

The results of the analysis of high-volume filters for brominated organics are presented in Tables IX and X. The interpretation of a positive finding is complicated by the possibility that the compounds found may not be from current emissions. They may be adsorbed onto soil particles which continue to be resuspended in air. The primary counter-indication of this explanation is the presence of varying combinations of compounds detected from day to day, especially TRIS and Firemaster 680, with very similar meteorology.

Conclusions

Hi-volume sampling with glass fiber filters may be used to collect and quantitate TRIS and Decabrom in the air. Qualitative and semi-quantitative information on other compounds such as Tetrabrom and Firemaster 680 may also be obtained. The analytical method used depends upon the compound being examined. Although glc/ms is the most widely applicable, it is not necessarily the method of choice. In the case of TRIS, the sensitivity using glc/ms/comp is very low when compared to glc/ecd. Although tlc/sd was not sensitive enough for the environmental samples encountered here it is specific for bromine compounds and could find application for screening in laboratories with limited instrumentation.

Field samples collected at the production facilities of these brominated compounds were analyzed by the procedures described here and reveal detectable quantities of each of these compounds.

Table IX. Brominated Organics Found on Glass Fiber Filters from Hi-Vol Samplers at Company A.

Date	Tetrabromobisphenol A (ng/m^3)	Firemaster 680 (ng/m^3)	Tris(2,3-dibromopropyl)phosphate (ng/m^3)	Decabromobiphenyl ether (ng/m^3)
12/19/76	—[a]	—	N.D.[b]	—
	—	—	N.D.[c]	—
12/21/76	—	—	N.D.[b]	—
	—	—	N.D.[c]	—
12/22/76	—	—	85 ± 8.6[b,d]	—
	—	—	53 ± 15[c,d]	—
7/22/77	28[e,f]	39[e,g]	60[d]	N.D.[e]
7/23/77	N.D.[e]	N.D.[e]	44[d]	N.D.[e]
7/24/77	N.D.[e]	172[e,g]	51[d]	N.D.[e]
7-25-77	N.D.[e]	183[e,g]	N.D.[d]	72[e,g]

[a]No indentification attempted.
[b]Glass fiber filter.
[c]Cellulose filter.
[d]Analysis by GLC/ECD. Results are uncorrected for recovery which was 87% at an equivalent to 27 ng/m^3 and 91% at an equivalent to 286 ng/m^3 of TRIS.
[e]GLC/MS/COMP analysis in the multiple ion mode.
[f]Ion intensity ratio of m/e 529/531 was 1.42 vs 1.48 for standard tetrabromobisphenol A.
[g]One ion monitored no confirmation.
N.D.= not detected.

Abstract

A method was developed and tested for collecting semi- and non-volatile brominated organic compounds from air using a glass fiber filter and a high-volume air sampler. Exposed filters were extracted with acetone and the extracts analyzed by either glc/ms/comp, glc/ecd or tlc. Recoveries of selected compounds from the filter material were >87%.

Ambient air sampling on plant property of industrial bromine extraction and bromo-organics synthesis facilities revealed the presence of tetrabromobisphenol A (N.D. −28 ng/m^3), decabromobiphenyl ether (N.D. −72 ng/m^3), tris(2,3-dibromopropyl)phosphate (N.D. −60 ng/m^3) and 1,2-bis(tribromophenoxy)ethane (N.D. −183 ng/m^3).

Table X. Results of the Analysis of Hi-Vol Filters for Semi-Volatile Brominated Organics — Company B.

Date	Tetrabromobisphenol A[b] (ng/m^3)	Decabromobiphenyl ether (ng/m^3)
12/13/76	Qual.[a]	<16[c,e] <13[d,e]
12/16/76	—	23[c,e] 94[d,e]
12/17/76	—	113[c,e] 118[d,e]
7/18/77	80[f]	8,000[f]
7/19/77	1,200[f]	1,000[f]
7/20/77	1,800[f,g]	25,000[f,h]
7/21/77	N.D.[f]	2,000[f]

[a]Not quantitated
[b]2,2-Bis(dibromo-4-hydroxyphenyl)propane.
[c]Cellulose filter
[d]Glass fiber filter
[e]Identified by TLC; confirmed and quantiated by GLC/MS/COMP
[f]Quantitation by gas chromatography-mass spectrometry with multiple ion detection.
[g]Confirmed by ion intensity ratio of 1.7 for the samples vs 1.5 for the standard for 529/531.
[h]Confirmed by ion intensity ratio for the sample of 2.8 compound to 2.7 for the sample for m/e pair 800/804.

Acknowlegements

The authors wish to thank C. A. Billings, R. P. Cepko, B. H. Garber and R. B. Keefe (RTI) for their assistance with the field sampling. The mass spectral analysis by J. T. Bursey and L. Kelner (RTI) is gratefully acknowledged. Valuable assistance and discussions were also obtained from F. Hall, Environmental Protection Agency Region VI, Dallas, TX; J. Southall and staff, State of Arkansas Department of Pollution and Ecology, V. J. Decarlo and G. E. Parris, Environmental Protection Agency, Office of Toxic Substances, Washington, D. C. This Research was supported by EPA Contract No. 68-01-1978.

Literature Cited

1. Gutenmann, W. H. and Lisk, D. J., Bull. of Environ. Contam. and Tox., *14*, 61 (1975).

2. *Bioassay of Tris(2,3-dibromopropyl)phosphate for Possible Carcinogenicity*, CAS No. 126-72-7, NCI-CG-TR-76, DHEW Publication No. (NIH) 78-1326 (1978).

3. Pellizzari, E. D., "Analysis of Organic Air Pollutants by Gas Chromatography/Mass Spectroscopy," EPA 600/2-77-100, (June, 1977).

4. Bursey, J., Smith, D., Bunch, J., Williams, N., Berkely, R. and Pellizzari, E., Amer Lab., *9*, 35-4 (1977).

5. Pellizzari, E. D., Castillo, N. P., Willis, S., Smith, D., and Bursey, J. T., Div. of Fuel Chem., Preprint, Am. Chem. Soc.; *23* (No. 2) 144-155 (1978).

6. Erickson, M. D., Michael, L. C., Zweidinger, R. A., and Pellizzari, E. D., Environ. Sci. Technol., *12*, 927 (1978).

7. Pellizzari, E. D., "Identification of Components of Energy-Related Wastes and Effluents," EPA-600/2-78-004 (1978).

8. Pellizzari, E. D., Zweidinger, R. A., and Erickson, M. D., "Environmental Monitoring Near Industrial Sites: Brominated Chemicals," EPA-560/6-78-002 (1978).

9. Hahn, F., Mikrochemie, L., *17*, 228 (1935).

RECEIVED November 17, 1978.

Ion Chromatographic Analysis of Trace Ions in Environmental Samples

W. E. RICH and R. A. WETZEL

Dionex Corporation, Sunnyvale, CA 94086

Ion Chromatography (IC) is a new ion-exchange, liquid chromatographic technique which was developed by H. Small, T. Stevens and W. Bauman (1) of the Dow Chemical Co. This technique overcomes the problems of separation and detection of highly acidic or basic ions. Thus, the power of high speed liquid chromatographic separation can now be applied to analytical problems involving these types of ions. A number of these species are of concern to the EPA because they are environmental pollutants. In general, IC can be used to separate and detect ions whose dissociation constant (pK) is less than 7.

Because of its exceptional selectivity, sensitivity and speed, IC is particularly suited to applications involving analysis of anions and cations in wastewater, natural waters, source effluents, workplace environments, ambient air and rain water. The analysis of organic as well as inorganic ions can be performed by IC. Table I is a growing list of ions which have been successfully separated and detected. The principles of IC and selected applications to environmental pollutants are described in this paper.

Principles of Ion Chromatography

A detailed description of IC is given in reference 1; however, the basic principles of the method can best be described by an example. Figure 1 schematically represents both an anion and a cation IC analysis. In both cases, the instrumentation involves a pumping system, an eluent, an injection valve, an ion-exchange separator column, an ion-exchange suppressor column and a conductivity cell. The sample is first injected into the flow system; then the well known reaction equilibrium shown in Figure 1 results in the separation of sample anions or cations on the separator column (2).

The detection system for this ion-exchange liquid chromatographic separation includes a second ion-exchange column and a conductivity cell placed in series with the separator column.

The second column is called a suppressor column. Its function is to convert the eluent to a less conductive species while converting sample ions to a common form. This system enables conductimetric detection of the sample ions in a low conductivity background. The ion-exchange suppressor reactions are also shown in Figure 1. In the case of anion analysis, sodium carbonate and/or bicarbonate eluent is converted to a weakly conductive dilute carbonic acid while the sample ions are converted to strong-

0-8412-0480-2/79/47-094-233$05.00/0

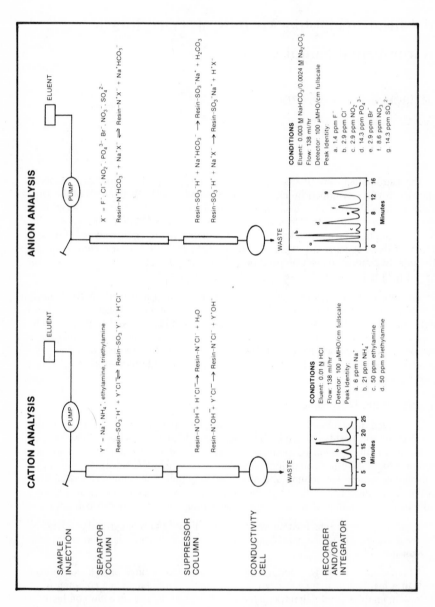

Figure 1.   Schematic of anion and cation IC analysis

TABLE I. A Growing List of Ions Separated and Detected By IC

Inorganic Ions

Ammonium	Chromate	Potassium
Arsenate	Dithionate	Rhenate
Azide	Fluoride	Rubidium
Barium	Hypochlorite	Sodium
Bromate	Iodate	Strontium
Bromide	Iodide	Sulfate
Calcium	Lithium	Sulfite
Carbonate	Magnesium	Tetrafluoroborate
Cesium	Nitrate	Thiocyanate
Chlorate	Nitrite	Thiosulfate
Chloride	Ortho-phosphate	

Organic Ions

Acetate	Ethyl methyl phosphonate	N-butylamine
Acrylate	Formate	Oxalate
Ascorbate	Fumarate	Propionate
Benzoate	Gluconate	Sarcosine
Butyrate	Glycolate	Succinate
Butyl phosphate	Hydroxycitrate	Tartarate
Citrate	Itaconate	Tetraethyl ammonium bromide
C1 Acetate	Lactate	Tetramethyl ammonium bromide
C1$_2$ Acetate	Maleate	Triethanolamine
C1$_3$ Acetate	Malonate	Triethylamine
Cyclohexylamine	Methacrylate	Trifluoro methane sulfonate
Dibutyl phosphate	Methyl phosphonate	Triisopropanolamine
Diethanolamine	Monoethylamine	Trimethylamine
Diethylamine	Monomethylamine	Tri-n-butylamine
Diisopropanolamine	Monisopropanolamine	
Dimethylamine		

ly conducting acid forms. In the case of cation analysis, the HCl eluent is converted to water while the sample ions are converted to strongly conducting basic forms. The conductimetric response is recorded by a strip chart recorder and/or electronic integrator. Comparison of retention time and peak height (or area) with those of standard solutions enables sample species identification and quantitation respectively. The linear response range extends from a few ppb to approximately 50 to 100 ppm (using a 100 μl injection volume) depending on the dissociation of the species. Above 100 ppm, conductivity is a simple function of concentration.

There are several important analytical consequences which can be realized from careful examination of Figure 1.

(1) H^+X^- and Y^+OH^- must be highly ionic (pK < 7) in order to have sensitive detection. Because ions with pK > 7 are **not** detected by IC, highly resolved chromatograms can be obtained from very complex mixtures.

(2) H^+X^- and Y^+OH^- must be stable species in solution. Amphoteric species such as amino acids are retained on the suppressor column and, therefore,

are not detected by IC. Also, cation transition metals generally precipitate as hydroxides on the OH-form suppressor resin and are not detected by IC.

(3) H^+ and Y^+OH^- must be species with less than about 20 carbon atoms. Larger organic species tend to absorb in the suppressor column resulting in excessive peak tailing or total absorption.

Ion Chromatography Applications

Any new analytical technique must be compared to existing techniques in order to determine its usefulness. IC has several characteristics which cause it to be the method of choice for several types of analyses. Some of the unique analytical capabilities of IC are:

(1) Multi-ion analysis per sample.

(2) Routine detection limits at low ppb concentrations.

(3) Separation from interfering ions which may limit accuracy of current analytical methods.

(4) Direct analysis of species which may require laborious pretreatment to enable separation or detection by current analytical methods.

(5) Ability to perform on-line analysis.

Because Ion Chromatography is a new technique, very little data has appeared in the literature describing its application to environmental analysis. The following discussion reviews what has been published and presents data on a number of new environmental applications.

A. Analysis of Wastewater and Natural Waters. The presence of certain anions in wastewater effluents can cause deterioration of natural water systems. Phosphorous and nitrogen can be present in several chemical forms in wastewaters. Phosphorous is usually present as phosphate, polyphosphate and organically-bound phosphorus. The nitrogen compounds of interest in wastewater characterization are ammonia, nitrite, nitrate and organic nitrogen. Analyses are often based on titrimetric, and colorimetric methods (3). These methods are time consuming and subject to a number of interferences. Ion Chromatography can be used to determine low ppm concentrations of these ions in less than thirty minutes with no sample preparation.

The use of IC for anion wastewater characterization is depicted in Figure 2. The separation of anions in wastewater from a process plant was performed using a standard eluent, 0.003 M $NaHCO_3$/0.0024 M Na_2CO_3. Total analysis time was approximately 24 minutes. Two unidentified peaks are present, one elutes between F^- and Cl^- and the second is partially resolved from the large Cl^- peak. The second unknown may be CO_3^{-2}, as exposure of the sample to air significantly reduced the response. NO_2^- may also be present and would be unresolved from the CO_3^{-2} under these conditions.

Carlson, et al (4) and C. Anderson, et al (5) have shown IC to be an excellent technique for low ppb analysis of water. Environmental applications for these analyses include determination of rain water composition and analysis of fuel cell effluents because ions in these solutions are often present in low ppb concentrations. Concentrator columns (3 x 50 mm) are used to accumulate sample ions to detectable quantities as large sample volumes are pumped through them. Samples may vary from 1 to 100

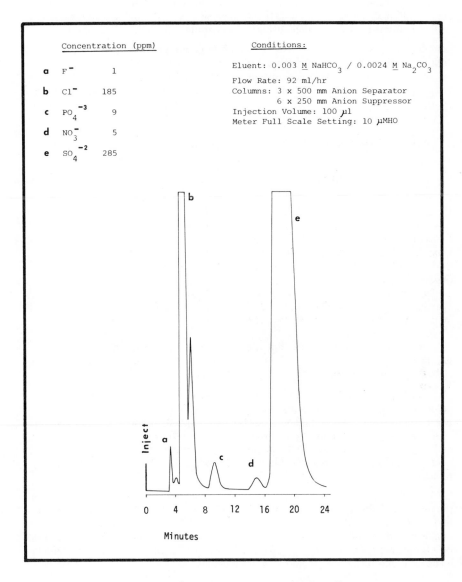

Figure 2.   *Analysis of anions in a wastewater effluent*

mls and are loaded by syringe or by pump. The concentrator then replaces the IC sample injection loop and routine analysis is performed. Due to the relatively unreactive location of sample ions on the concentrator resin, samples may be loaded remotely from the IC and stored several days before analysis. Figure 3 shows the chromatogram of low ppb levels of four anions using a concentrator and a 10 ml sample volume.

Ion Chromatography has also been used to determine the quality of city water supplies from ten different locations during the summer of 1977 (6). Chromatograms are shown for city water supplies collected from four cities (Figure 4). Table II summarizes the data obtained for all ten cities with a total analysis of less than four hours.

TABLE II. Analysis of City Water Supplies Utilizing Ion Chromatography

	pH	F^-	Cl^-	NO_2^-	NO_3^-	SO_4^{-2}	PO_4^{-3}
					Concentration (ppm)		
Atlanta, GA	7.40	0.32	5.7	0	<0.5	6	<1
Dearborn, MI	8.15	0.63	9.3	0	<0.5	19	<1
Los Angeles, CA	8.12	0.21	75.0	0	17	146	2.7
Louisville, KY	7.86	1.05	34.3	0	3.4	81	<1
Oakville, ONT.	7.72	0.60	33.6	0	<0.5	30	1.0
St. Louis, MO	9.47	0.76	21.6	0	3.3	132	<1
St. Thomas, ONT.	8.47	0.56	21.6	0	<0.5	27	<1
St. Paul, MN	8.08	—	20.0	0	<0.5	32	<1
Wayne, MI	8.35	0.54	10.2	0	1.4	18	<1
Wixom, MI	8.13	0.54	22.8	0	0.9	22	<1

B. Analysis of Atmospheric Emissions from Stationary Sources. Ion Chromatography is applicable to the analysis of gaseous and particulate species in vehicle and stationary source emissions.

An Ion Chromatograph has been successfully automated by interfacing it to an automatic sampler (7). Continuous unattended analysis was possible, the actual number of samples analyzed being limited by the ionic capacity of the suppressor column. The automated Ion Chromatograph was used to analyze soluble sulfates, ammonia and alkyl amines in stack and automobile exhaust samples. Excellent agreement between IC and automated barium chloroanilate titration for sulfate was obtained with a relative standard deviation less than 5%.

An application in power production, particularly in coal-fired power plants, is the analysis of flue gas scrubbers which remove excess SO_2 following coal combustion. Tests run by SAMBESRL at the EPA's Research Triangle Park facility (8,9) have demonstrated the effectiveness of IC in determining sulfite and sulfate in flue gas desulfurization systems. Table III gives results of direct IC analysis of scrubber liquors compared with turbidimetric and titration methods.

C. Ambient Air Analysis. Trace analysis of toxic gases in the environment is receiving more emphasis by many industries. Some of the more polar gases are very difficult to analyze by gas chromatography. One example is monochloroacetyl chloride. This gas can be collected from the atmosphere on Silica gel, extracted in dilute sodium bicarbonate, and then analyzed directly by IC as monochloroacetic acid (10). A

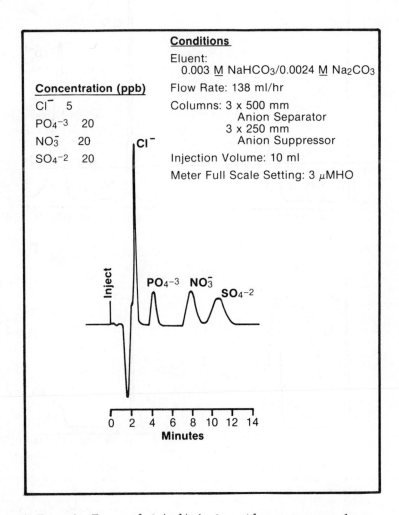

Conditions

Eluent:
0.003 $\underline{M}$ NaHCO$_3$/0.0024 $\underline{M}$ Na$_2$CO$_3$

Flow Rate: 138 ml/hr

Columns: 3 x 500 mm
Anion Separator
3 x 250 mm
Anion Suppressor

Injection Volume: 10 ml

Meter Full Scale Setting: 3 μMHO

Concentration (ppb)

Cl$^-$ 5
PO$_4^{-3}$ 20
NO$_3^-$ 20
SO$_4^{-2}$ 20

Inject

Cl$^-$

PO$_4^{-3}$ NO$_3^-$
SO$_4^{-2}$

0 2 4 6 8 10 12 14
Minutes

Figure 3.   Trace analysis (ppb) of anions with a concentrator column

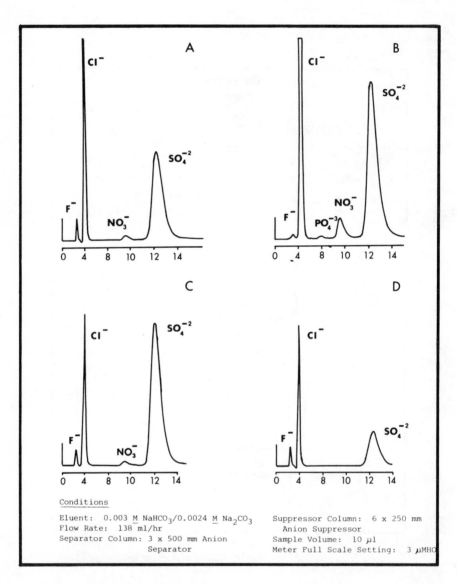

Figure 4. Ion chromatograms of drinking water samples

TABLE III. Analysis of Flue Gas Desulfurization Liquor*

Ion Chromatography		Total SO$_x$ as SO$_4^{-2}$	Ba^{+2} titration Total SO$_x$ as SO$_4^{-2}$	Turbidimetric Total SO$_x$ as SO$_4^{-2}$
SO$_3^{-2}$	SO$_4^{-2}$			
2.0	2.1	4.5	4.4	4.2
1.5	2.7	4.5	3.7	4.0
0.0	2.4	2.4	2.5	2.5
0.0	2.1	2.1	2.0	1.7
0.7	3.3	4.1	4.1	4.1
0.0	1.4	1.4	1.5	1.3
0.8	1.8	2.7	2.6	2.7
0.0	1.1	1.1	1.2	1.1

*Results are given in mg/ml

chromatogram of monochloroacetic acid, acetic acid and hydrochloric acid is shown in Figure 5.

Another gas of environmental interest is SO$_2$. Collection of ambient air containing SO$_2$ in a dilute H$_2$O$_2$ scrubber quantitatively converts SO$_2$ to SO$_4^{-2}$. The scrubber solution can then be injected directly into an IC and SO$_2$ is determined as SO$_4^{-2}$ (9). This method has the advantage over other methods (e.g. West-Gaeke) in that the collection solution is temperature stable.

Sulfur dioxide and nitrogen oxides present in ambient air can be partially trapped as aerosols in the atmosphere. Because the acid forms of these sulfate and nitrate particulates can be corrosive when removed from the atmosphere by rainwater, they are currently being monitored by the U.S. EPA (9,11). Their toxicity to human beings is also being studied by the EPA. Mulik, et al (11) have shown IC to be excellent for the analysis of nitrate and sulfate in ambient aerosols. An air sample is collected through high volume filters. The filters are then extracted with water or dilute carbonate; and the extract is directly injected into the IC. Results, including comparison to standard methods, are listed in Table IV. A more detailed comparison with standard methods is given in (9,11).

Ion Chromatography has also been utilized for the determination of azide in water effluents and in particulates generated from the deployment of an air bag system. In a recent study (12) particulate material was collected and leached with deionized water. The solutions were then analyzed by Ion Chromatography for anions. A typical chromatogram showing acetate, chloride, nitrite, azide and sulfate is shown in Figure 6. The concentration of azide was also analyzed using a colorimetric technique. Results for equivalent samples correlated to within ± 5% (12).

D. Rainwater Analysis. IC is also being applied to the analysis of ions in rainwater (13). Measurements of precipitation samples are being collected at remote or baseline stations to provide an estimate of the natural ion concentration observed in rain unaffected by man's activity. These measurements will provide data, to study increases of certain constituents with time due to energy and industrial production.

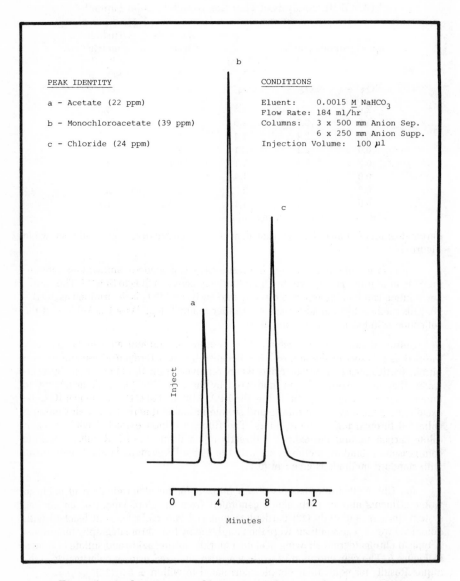

PEAK IDENTITY

a - Acetate (22 ppm)

b - Monochloroacetate (39 ppm)

c - Chloride (24 ppm)

CONDITIONS

Eluent: 0.0015 M NaHCO₃
Flow Rate: 184 ml/hr
Columns: 3 x 500 mm Anion Sep.
             6 x 250 mm Anion Supp.
Injection Volume: 100 μl

Figure 5.   Analysis for monochloroacetic acid in an ambient atmosphere

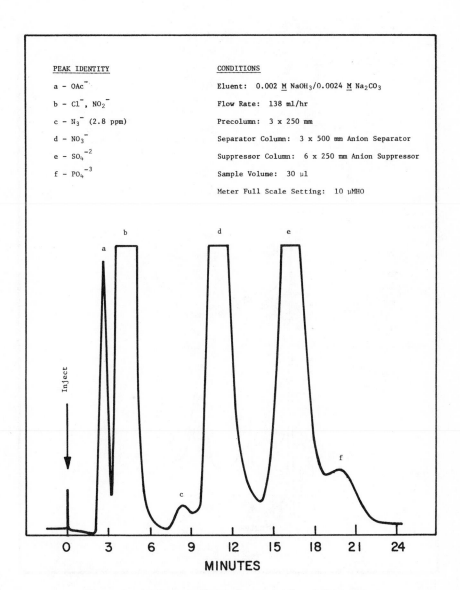

Figure 6.   Analysis of effluents from an air bag inflator system

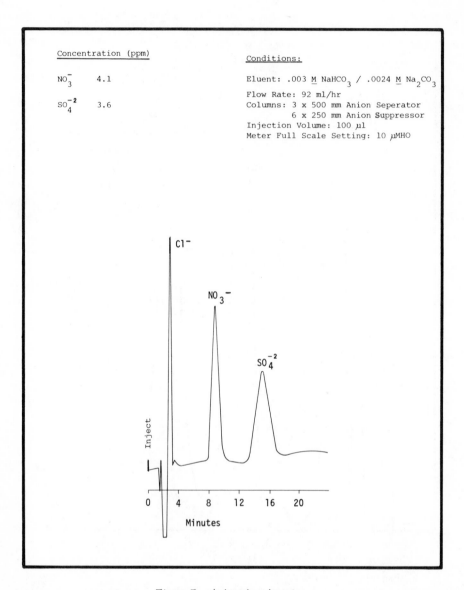

Figure 7.   Anions in rainwater

TABLE IV. Sulfate and Nitrate Analysis

| Sample Number | $SO_4^{-2}(\mu g/ml)$ | | $NO_3^-(\mu g/ml)$ | |
	IC	Methyl thymol blue	IC	Cadmium reduction
1	65.1	66.2	9.23	11.9
3	79.4	72.9	21.2	20.4
5	29.2	30.9	6.4	7.0
7	52.9	53.2	23.6	23.5
9	35.5	35.3	9.3	9.6
11	88.8	82.1	4.3	4.5
13	23.6	25.1	5.2	5.9
15	35.9	36.5	17.5	18.2
17	47.6	48.7	12.3	12.4

The samples are analyzed for Na^+, K^+, NH_4^+, Mg^{+2}, Ca^{+2}, Cl^-, NO_3^-, SO_4^{-2} and PO_4^{-3}. Results of sample analyses by IC were compared with results obtained by other methods such as atomic absorption and spectrophotometry. An agreement of ± 10% was found at concentrations from 50-5000 $\mu g/l$ and interlaboratory comparisons showed agreements of ± 15%. Figure 7 shows an anion chromatogram of rainwater.

Conclusions

Ion Chromatography is a new and sensitive analytical technique which provides the efficiency of high performance liquid chromatography to chemists dealing with problems involving highly acidic and basic ions. It has been shown that this technique has the capability of analyzing for a number of toxicologically important species in environmental samples. This system has the advantage that it operates in a continuous mode and can perform an analysis in two or three minutes per ion. In conclusion, it is expected that Ion Chromatography will replace a number of classical techniques currently used for the analysis of ionic species in environmental samples.

Literature Cited

1. Small, H., Stevens, T. S., Bauman, W. C., Anal. Chem., 47, (11), 1801 (1975).

2. Rieman, W., III, Walton, H. F., "Ion Exchange in Analytical Chemistry," published by Piergamon, 1970, Elmsford, N.Y.

3. *Standard Methods for the Examination of Water and Waste Water*, 14th Edition, American Public Health Association (1975).

4. Carlson, G. L., et al, "Rapid Analysis of Boiler Water to ppb Level by Ion Chromatography," paper #362, Pittsburgh Conference on Analytical Chemistry and Applied Spectroscopy, Cleveland, Ohio (1978).

5. Anderson, C., Schleicher, H., "Rapid Analysis of Water to ppb Level by Ion Chromatography," paper #363, Pittsburgh Conference on Analytical Chemistry and Applied Spectroscopy, Cleveland, Ohio (1978).

6. Westwood, L., Schuetzle, D., Ford Research Laboratory, private communication.

7. Zweidinger, R. B., Tejada, S. B., Sigsby, J. E., Bradow, R. L., "Application of Ion Chromatography to the Analysis of Ammonia and Alkylamines in Automobile

Exhaust." *Ion Chromatographic Analysis of Environmental Pollutants.* Edited by Sawicki, E., Mulik, J. D., Wittgenstein, E., Ann Arbor Science Publishers, Ann Arbor, MI (1978).

8. Steiber, R., Statnick, R. M., "Applications of Ion Chromatography to Stationary Source and Control Device Evaluation Studies." *Ion Chromatographic Analysis of Environmental Pollutants.* Edited by Sawicki, E., Mulik, J. D., Wittgenstein, E., Ann Arbor Science Publishers, Ann Arbor, MI (1978).

9. Mulik, J. D., Todd, G., Estes, E., Puckett, R., Sawicki, E., "Ion Chromatographic Determination of Atmospheric Sulfur Dioxide." *Ion Chromatographic Analysis of Environmental Pollutants.* Edited by Sawicki, E., Mulik, J. D., Wittgenstein, E., Ann Arbor Science Publishers, Ann Arbor, MI (1978).

10. Menear, J., private communication (1978).

11. Mulik, J., Puckett, R., Williams, D., Sawicki, E., "Analysis of Nitrate and Sulfate in Ambient Aerosols," Anal. Letters 9 (7): 653 (1976).

12. Westwood, J., and Stokes, E. L., "Analysis of Azide in Environmental Samples by Ion Chromatography." Submitted to Analytical Letters (June, 1978).

13. Bogen, D. C., et al, "Baseline Precipitation Chemistry Measurements by Ion Chromatography," paper #361, Pittsburgh Conference on Analytical Chemistry and Applied Spectroscopy, Cleveland, Ohio (1978).

RECEIVED November 17, 1978.

N-Nitroso Compounds in the Workplace

DAVID H. FINE

Cancer Research Division, Thermo Electron Corporation,
45 First Avenue, Waltham, MA 02154

N-nitroso compounds are widely recognized as being unquestionably nasty materials. More than 100 of the 130 different N-nitroso compounds tested in animals have been shown to be carcinogenic (1). The site of the tumor depends upon the chemical structure of the N-nitroso compound, the route of administration and the species being tested. Three N-nitroso compounds, N-nitrosodimethylamine (NDMA), N-nitrosodiethylamine (NDEA) and N-nitrosopyrrolidine (NPYR), have been tested in dose response studies with rats; between 1 and 5 parts per million (ppm) of these nitrosamines in the diet was found to produce a marginally significant number of tumors in a rat population of under 100 animals (2, 3, 4).

As recently as two years ago it was generally believed that human exposure to N-nitroso compounds was limited to at most a few micrograms (μg) per day of NDMA and NPYR in nitrite preserved foodstuffs and similar amounts of N-nitroso nornicotine (NNN) in tobacco smoke. During the past several years this delusion has crumbled as significant levels of N-nitrosoamines have been found in air (5, 6), soil (7), water (8), pesticides (9), cutting fluids (10), cosmetics (11), tobacco (12) and even *in vivo* in man (13). Other major routes of human exposure probably still remain undiscovered. It is becoming increasingly apparent that N-nitroso compounds are ubiquitous in the environment, particularly in the many chemical, agricultural and consumer products which characterize our modern society.

N-nitroso compounds have the general formula:

$$\begin{array}{c} R_1 \\ \diagdown \\ \diagup \\ R_2 \end{array} N - NO$$

Where R_1 and R_2 can be virtually any organic group. For example, if R_1 and R_2 are alkyl groups, it is an alkyl N-nitrosamine; if R_1 or R_2 contains a carboxylic acid group, it is an N-nitrosamino acid; if R_1 is an alkyl group and R_2 is an amino group, it is an N-nitrosourea. N-nitroso compounds can be formed by the reaction of various precursor entities. The amine fragment, (R_1R_2N) can come from primary, secondary or tertiary amine. The nitrosyl group, NO, can be derived from nitrogen oxides (NO, NO_2, N_2O_4, or N_2O_3), nitrite (nitrite salts or nitrous acid) or certain C-nitro compounds. N-nitrosation of the amine fragment can also occur via transnitrosation by other, more labile, N-nitroso compounds. Depending on the reactants and the catalysts which are present, N-nitrosation can occur at either acidic, neutral or alkaline conditions. Some known

0-8412-0480-2/79/47-094-247$05.00/0
© 1979 American Chemical Society

N-nitrosation catalysts include formaldehyde, chloral, ozone, and metal ions.

CHEMISTRY

Primary amines. The nitrosation of primary amine type compounds yields small amounts of secondary amine type N-nitroso compounds (14, 15, 16). The reaction mechanism is not well understood.

Secondary amines. The reaction of secondary amine type compounds with nitrous acid (HO · NO) has been reviewed extensively by Turney and Wright (17), Ridd (18), Scanlan (19) and Mirvish (20). In a system containing (HO · NO) as the nitrosating agent, the possible nitrosyl carriers are (H_2O · NO), (NO_2 · NO) and (NO^+). The reactivity of (NO^+) is very low and it is not considered an effective nitrosating form. Nitrous acidium ion (H_2O · NO^+) plays a significant role only at concentrated acidic condi tions. Therefore, it seems likely that at the dilute acidic conditions that are encountered in the environment, it is nitrous anhydride (N_2O_3 = NO_2 · NO) which nitrosates secondary amines.

$$2 \ HNO_2 \overset{fast}{\rightarrow} NO_2 \cdot NO + H_2O$$

$$NO_2 \cdot NO + R_2NH \rightarrow R_2 \ N \cdot NO + NHO_2$$

For most secondary amines the second equation is rate limiting. The kinetic expressions is thus

$$Rate = k \ \{HNO_2\}^2 \ \{R_2NH\}$$

The reaction rate is pH dependent and has a maximum value at pH 3.4, which is the pKa of nitrous acid. The ease of nitrosation increases as the basicity of the amine decreases.

In the presence of the nitrosyl anion, (X · NO), where X is the anion, NO is an efficient nitrosating agent (17). The effectiveness of the nitrosating species is in the following order (18).

$$NCS \cdot NO = I \cdot NO > Br \cdot NO > Cl \cdot NO > NO_2 \cdot NO$$

The nitrosation of amides, such as methylurea and methylurethane, follows the kinetic expression:

$$Rate = k_2 \ (amide) \ (HNO_2) \ (H^+)$$

The main nitrosating agent is probably nitrous acidium ion ($NO \cdot OH_2^+$). There is no simple rule, as there is for secondary amines, relating the ease of nitrosation to the properties of the amide.

The N-nitrosation reaction is usually very slow at neutral or alkaline pH due to the low equilibrium concentration of anhydrous nitrous acid. However, in the presence of formaldehyde or chloral as a catalyst (21), appreciable nitrosation occurs, even at pH 6 to 11. Similarly, Keefer (22) showed that some metal ions could catalyze the reaction under basic conditions.

Tertiary amines. Tertiary amine type compounds, react with nitrous acid to yield secondary-amine type N-nitroso compounds. The myth that tertiary amines do not nitrosate to yield N-nitroso compounds, is a remarkable feat of misinformation that has persisted for over 100 years (23, 24, 25).

The mechanism for the N-nitrosation of tertiary amines assumes that the unshared electron pair on the unprotonated amine reacts with a nitrosating species to form the nitrosammonium ion (see Figure 1) (26, 27). The nitrosammonium ion then undergoes cis elimination of nitroxyl to form an immonium ion. At 100°C and low pH (pH = 3), the immonium ion is hydrolyzed to give a carbonyl compound and a protonated secondary amine; the protonated secondary amine is then nitrosated to the corresponding N-nitrosamine. At 100°C and higher pH (pH = 6) or at 25°C, the immonium ion undergoes nucleophilic attack by free nitrite, to form an unstable adduct which decomposes to form a carbonyl compound and the nitrosamine.

The reaction of tertiary amine oxides with nitrous acid has also been shown to produce N-nitroso compounds. The mechanism for the amine oxides is similar to that for the tertiary amines (26).

Nitrogen oxides. Nitric oxide (NO) itself, has been shown to be a poor nitrosating agent (28), probably because it is unable to abstract an amino-H atom to generate the dialkyl-amino radical, which might then combine with further NO. However, the presence of even a small amount of air results in complete conversion, presumably via oxidation of NO to NO_2. Nitrosation by NO is catalyzed by metal salts, such as ZnI_2, CuCl, and $CuSO_4$. The metal catalyzed reaction is inhibited in acid or aqueous media (29).

Nitrosation by NO and NO_2 proceeds rapidly in aqueous solution at a pH in the range 7-14. The nitrosating agent is presumably N_2O_3 and/or N_2O_4. However, the reaction rate for NO_2 and N_2O_3 at neutral and alkaline pH often exceeds that for nitrite nitrosation under acidic condition.

N-nitrosation by oxides of nitrogen at neutral and alkaline pH has an important bearing on assessing human exposure to N-nitroso compounds, particularly as this route has been almost totally disregarded in the past. Those analytical techniques which have relied on a basic pH to 'inhibit' nitrosation need further study to ensure the validity of the findings. For the same reason, measurements of N-nitrosamines in engine exhausts and tobacco smoke are likely to be particularly artifact prone.

C-nitro compounds. Fan et al, (30) have recently shown that certain C-nitro compounds can readily nitrosate secondary amines to form N-nitroso compounds. The reaction is independent of the basicity of the amine and occurs rapidly at neutral and alkaline pH. Certain structure requirements are necessary for C-nitro compounds to act as nitrosating agents. For alkyl C-nitro compounds, an electron withdrawing group attached to the same carbon atom as the nitro group is necessary. For aromatic C-nitro compounds, an electron withdrawing group on a suitable position on the aromatic ring is required. Because C-nitro compounds are widely used as pesticides, bactericides, coloring agents, drugs and perfumes, nitrosation by C-nitro compounds may be important.

ANALYSIS

Because N-nitroso compounds can have such a wide variety of physical and chemical properties, and because they can be formed from a wide variety of precursors, analysis at the trace level is difficult. The most widely used technique is the use of a nitrosamine specific detector, called a TEA, which can be interfaced to either a gas chromatograph (GC) or a high pressure liquid chromatograph (HPLC) (31, 32). General screening procedures which have been designed to detect all N-nitroso compounds have been developed (33, 34). Structural confirmation of N-nitroso compounds is generally by high resolution mass spectrometry (35). Extensive artifact experiments are necessary in order to ensure that the N-nitroso compound which was identified had not been formed during the analysis procedure itself (36).

Because N-nitroso compound analysis is so prone to artifacts, analysis of ambient air is best carried out using a mobile laboratory which contains the necessary analytical equipment (37). If positive results are obtained, extensive artifact experiments can then be carried out on site, in the presence of the likely precursors. After it has been demonstrated that N-nitroso compounds are being neither formed or lost during the analytical procedures, then further samples are collected for subsequent analysis by high resolution mass spectrometry. A dry pre-packaged cartridge, called Thermosorb™ (Thermo Electron Corp.,) has recently become available for airborne monitoring of N-nitroso compounds.

AIRBORNE N-NITROSO COMPOUNDS

The simplest dialkyl N-nitrosamines are relatively volatile, and airborne N-nitrosamines are to be expected in the vicinity of chemicals which are contaminated with nitrosamines. In situ, formation of N-nitrosamines would be expected to occur if amines were present in close proximity to nitrosating agents such as nitrite, oxides of nitrogen and C-nitro compounds.

Rocket fuel factory and surroundings. Fine, et al (5) reported nitrosodimethylamine (NDMA) to be present as an air pollutant in Baltimore, Maryland. The prime source was subsequently found to be a chemical plant which was manufacturing the rocket fuel, unsymmetrical dimethyl hydrazine (UDMH), for which NDMA was used as a precursor. Typical NDMA levels were between 6000 and 36,000 ng/m^3 on the site of the factory, about 1000 ng/m^3 in the residential neighborhood adjacent to the factory and about 100 ng/m^3 two miles away in downtown Baltimore. Typical daily human exposures to NDMA were calculated to be 39 μg at the factory site, 10 μg in the adjacent residential neighborhood and 1.0 μg in downtown Baltimore. It is of interest to note that a leak of just 130 g. (4.7 ounces) NDMA per hour could have explained all the airborne NDMA emissions in Baltimore.

Amine factories. Bretschneider and Matz (38) reported NDMA to be present in the air at trace levels in a factory producing fat chemicals, and in a pharmaceutical plant. They also reported levels of between 1000 and 43,000 ng/m^3 on the site of a high-pressure dimethylamine synthesis plant. Fine, et al (5) reported NDMA to be present in the ambient air at the 1-100 ng/m^3 level outside a factory in Belle, West Virginia which was manufacturing and using dimethylamine. Subsequent studies (39) showed that the source of the NDMA was a small lots vent (levels of up to 130,000 ng/m^3) and a small lots waste car (levels of up to 3,000,000 ng/m^3), where the NDMA was apparently being produced as an unwanted by-product. The NDMA levels in the neighboring towns of Belle and Charleston were in the 1-40 ng/m^3 range, no evidence was found to suggest that significant amounts of NDMA were being formed by atmospheric reaction of dimethylamine with oxides of nitrogen.

Rubber and Tire Industry. Fajan et al (40) showed that N-nitrosomorpholine, and to a lesser extent, NDMA, were present as air pollutants inside a tire factory and a factory where chemicals for the tire industry were being manufactured. The N-nitrosomorpholine levels varied between 600 and 7,200 ng/m^3. The results may be relevant to recent epidemiological studies (41) which report an increased risk of certain types of cancer in workers in the same areas of the tire factories where we found the highest N-nitrosomorpholine levels.

Leather Tanneries. N-nitrosodimethylamine has recently been shown *(40)* to be present as an air pollutant inside a leather tannery, at levels between 100 ng/m^3 and 47,000 ng/m^3. The average NDMA level in the 19 air samples where were collected was 17,000 ng/m^3. This is the highest known human exposure to N-nitrosamines.

Ambient Air. Hanst, *et al (42)* showed that dimethylamine could react with nitrogen oxides in the atmosphere to form N-nitrosodimethylamine. Fine, *et al (39)* in studies in New York City, Boston and upstate New Jersey, found little evidence to suggest that NDMA or other N-nitroso compounds were being formed in the atmosphere, even in the neighborhood of amine factories. NDMA was found sporadically, in only 3 out of 40 sites.

Tobacco Smoke. The measurement of N-nitroso compounds in tobacco smoke is a difficult task, since the smoke contains relatively large amounts of precursor amines, as well as nitrogen oxide levels as high as 1000 ppm *(14)*. Aging of the smoke, either in traps or in collection bags can lead to significant artifact formation. This problem has been avoided by Hoffmann and co-workers *(12, 43)* who have reported NDMA, N-nitrosomethylethylnitrosamine (NMEA), NDEA, NPYR and NNN to be present in both mainstream smoke (smoke which is sucked into the mouth) and sidestream smoke (the smoke which escapes into the surroundings). Typical levels in the mainstream smoke are 5.7-4.3 ng NDMA/cigarette, 5.1-22 ng NPYR/cigarette and 137 ng NNN/cigarette. Sidestream smoke contains between 20 and 200 times as much nitrosamine as the mainstream smoke. The amount of nitrosamine in the sidestream smoke depends on such factors as the air flow rate in the neighborhood of the burning cone, and the age of the smoke.

Indoor NDMA pollution of air from the burning of tobacco has been investigated *(43)*. Typical NDMA concentrations were 240 ng/m^3 in a New York bar, 130 ng/m^3 in the club car of a commuter train, 90 ng/m^3 in a sports hall and a discotheque and non-detectable in a private residence.

Vapors from cooking. About 90% of NDMA and 75% of NPYR formed in cooked bacon is lost through evaporation during frying *(44, 45)*. It has been calculated that the airborne nitrosamine levels in a typical kitchen could be as high as 56 ng/m^3 of NDMA and 560 ng/m^3 of NPYR *(16)*. However, the exposure would be of short duration (<1 hr.) and the total nitrosamine intake would be small.

Miscellaneous. Volatile nitrosamines can escape into the atmosphere from a variety of other sources. Automobile and diesel engine exhausts may contain N-nitroso compounds, including NDMA at trace levels *(5)*. Nitrosodiethanolamine (NDE1A) is a likely air contaminant in machine shops which use cutting and grinding fluids contaminated with high concentrations of NDE1A *(10)*. Several herbicides, known to contain appreciable levels of volatile nitrosamines *(9)*, are applied as aqueous sprays; it is likely that worker exposure via inhalation may be appreciable.

ACKNOWLEDGMENTS

It is a pleasure to acknowledge many useful discussions with Gordon Edwards, Tsai Y. Fan, Ira Krull, David Rounbehler and Martin Wolf. This work was supported by the National Science Foundation (NSF) under Grant No. ENV75-20802, The National Institute of Occupational Safety and Health (NIOSH) under Contract 210-77-001 and the Environmental Protection Agency (EPA) under Contract No. 68-02-2766. Any opinions, findings, conclusions and recommendations expressed are those of the authors and do not necessarily reflect the views of NSF, NIOSH or EPA.

REFERENCES

1. Magee, P. N., R. Montesano, and R. Preussmann, *Chemical Carcinogens*, C. E. Searle (ed) ACS Monograph #173, American Chemical Society, Washington, D.C.

2. Terracine, B., P. N. Magee and J. M. Barnes, *Br. Journal of Cancer*, (1967) *21*, 559.

3. Druckrey, H., A. Schilbach, D. Schmahl, R. Preussmann and S. Ivankovic, *Arzneimittelforschung.* (1963) *13*, 841.

4. Preussmann, R., D. Schmahl, G. Eisenbrand and R. Port, *Proceedings of the Second International Symposium on Nitrite in Meat Products*, Zeist, The Netherlands, B. J. Tinbergen and B. Krol (eds) Pudoc, Wageningen Centre for Agricultural Publishing and Documentation, p. 261.

5. Fine, D. H., D. P. Rounbehler, N. M. Belcher and S. S. Epstein, *Science,* (1976) *192*, 1328.

6. Fine, D. H., D. P. Rounbehler, E. D. Pellizzari, J. E. Bunch, R. W. Berkley, J. McCrae, J. T. Bursey, E. Sawicki, K. Krost and G. A. Demarrais. *Bull. Env. Contam. & Tox. 15*, 639 (1976).

7. Fine, D. H., D. P. Rounbehler, A. Rounbehler, A. Silvergleid, E. Sawicki, K. Krost and G. A. DeMarrais, *Environ. Sci. & Tech.*, (1977) *11*, 581.

8. Fine, D. H. and D. P. Rounbehler, *Identification and Analysis of Organic Pollutants in Water*, L. H. Keith (ed) Ann Arbor Science, 1976.

9. Ross, R., J. Morrison, D. P. Rounbehler, S. Y. Fan and D. H. Fine, *J. Agr. Fd. Chem.*, (1977) *25*, 1416.

10. Fan, T. Y., J. M. Morrison, D. P. Rounbehler, R. Ross, D. H. Fine, W. Miles and N. P. Sen, *Science*, (1977) *196*, 70.

11. Fan, T. Y., U. Goff, L. Song, D. H. Fine, G. P. Arsenault, and K. Biemann, *Food and Cosmetics Toxicol.*, (1977) *15*, 423.

12. Hoffmann, D., S. S. Hecht, R. M. Ornaf, T. C. Tso, and E. L. *Environmental N-Nitroso Compounds Analysis and Formation*, E. A. Walker, P. Bogovski and L. Gricuite (eds) IARC Scientific Pub. #14, p. 307, Lyon, France (1976).

13. Fine, D. H., R. Ross, D. P. Rounbehler, and L. Song, *Nature*, (1977) *265*, 753.

14. Fine, D. H. "N-nitroso Compounds in the Environment" in Pitts, J. (ed) *Advances in Environmental Science and Technology*, Vol. 9, Wiley Interscience (In press).

15. Wartheson, J. J., Ph. D. Thesis, Oregon State University, Corvallis, Oregon (1975).

16. Wartheson, J. J., R. A. Scanlan, D. P. Bills, and L. M. Libbey, *J. Agr. Fd. Chem.*,(1975) *23*, 898.

17. Turney, T. A. and G. A. Wright, *Chem. Rev.*, (1959) *59*, 497.

18. Ridd, J. H. *Quart. Rev. Chem. Soc. (London)*, (1961) *15*, 418.

19. Scanlan, R. A., *C.R.C. Critical Review in Food Technology*, (1975) 357.

20. Mirvish, S. S. *Toxicol. Appl. Pharm.*, *31*, (1975) 325.

21. Keefer, L. K. and P. P. Roller, *Science*, (1973) *181*, 1245.

22. Keefer, L. K. *Environmental N-Nitroso Compounds Analysis and Formation*, E. A. Walker, P. Bogovski and L. Griciute (eds) IARC Scientific Pub. #14, p. 513, Lyon, France.

23. Geuther, A., *Arch. Pharm. (Weinheim)*, (1864) *123*, 200.

24. Walters, C. L., *Chemistry in Britain*, (1977) *13*, 140.

25. Hein, G. E., *J. Chem. Educ.*, (1963) *40*, 181.

26. Ohshima, H. and T. Kawabata, in *Environmental Aspects of N-Nitroso Compounds*, E. A. Walker, M. Castegnaro, L. Griciute and R. E. Lyle (eds) IARC Scientific Publication No. 19, Lyon, 1978.

27. Smith, P. A. S. and R. N. Loeppky, *J. Amer. Chem. Soc.*, (1967) *89*,1147.

28. Challis, B. C. and S. A. Kyrtopoulos, *J.C.S. Chem. Comm.*, (1976) *178*.

29. Challis, B. C., A. Edwards, R. R. Hunma, S. A. Kyrtopoulos and J. R. Outram, in *Environmental Aspects of N-Nitroso Compounds*, E. A. Walker, M. Castegnaro, L. Griciute and R. E. Lyle (eds) IARC Scientific Publication No. 19, Lyon, 1978.

30. Fan, T. Y., R. Vita and D. H. Fine, *Toxicology Letters* (1978) 2,5.

31. Fine, D. H. and D. P. Rounbehler, *J. Chrom.*, (1975) 109, 271.

32. Fine, D. H., D. P. Rounbehler, A. Silvergleid, and R. Ross, *Proceedings of the Second International Symposium on Nitrite* in Meat Products, B. J. Tinbergen and B. Krol (eds) Zeist, The Netherlands, 191 (1977).

33. Fan, T. Y., I. S. Krull, R. D. Ross, M. H. Wolf and D. H. Fine, in *Environmental Aspects of N-nitroso Compounds*, E. A. Walker, M. Castegnaro, L. Griciute and R. E. Lyle (eds) IARC Scientific Publication No. 19, Lyon, 1978.

34. Krull, I. S., T. Y. Fan, M. H. Wolf, R. D. Ross, and D. H. Fine, Liquid Chromatography Symposium I: Biological/Biomedical Applications of LC, Marcel Dekker, N.Y., 1978 (in press).

35. Gough, T. A., K. S. Webb, M. A. Pringuer and B. J. Wood, *J. Agr. Fd. Chem.* (1977) *25*, 663.

36. Krull, I. S., T. Y. Fan and D. H. Fine, *Analyt. Chem.* (1978) 50, 698.

37. Krull, I. S., T. Y. Fan, D. H. Fine, R. Young and E. Egan, *Research/Development Magazine*, May, 1978 (p. 117).

38. Bretschneider, K. and J. Matz. *Arch. Geschwulstforsch.* (1973) 42, 36.

39. Fine, D. H. *Toxic Substances in the Air Environment*, J. D. Spengler, (ed), Pittsburgh, APCA, 168 (1977).

40. Fajen, J. M., G. A. Carson, T. Fan, D. P. Rounbehler, J. Morrison, I. Krull, G. Edwards, A. Lafleur, W. Herbst, U. Goff, R. Vita, K. Mills, D. H. Fine and V. Reinhold, paper presented at Air Pollution Control Association Annual Meeting, Houston #78-6.7, June 26, 1978.

41. Monson, R. R. and K. K. Nakano, *Am. J. Epidemiol.* 103, 284 (1976); Peters, J. M., R. R. Monson, W. A. Burgess and L. J. Fine, *Env. Hlth. Perspectives*, (1976) 17, 31.

42. Hanst, P. L., J. W. Spence, F. M. Miller, paper presented at ACS, April, 1976.

43. Brunnemann, K. D. and D. Hoffmann, in Environmental Aspects of N-Nitroso Compounds, E. A. Walker, M. Castegnaro, L. Griciute and R. E. Lyle (eds) IARC Scientific Publication No. 19, Lyon, 1978.

44. Gough, T. A., K. Goodhead and C. L. Walters, *J. Sci. Fd. Agric.*, (1976) 27, 181.

45. Sen. N. P., S. Seaman, and W. F. Miles, *Fd. Cosmet. Toxicol.*, (1976) 14, 167.

RECEIVED November 17, 1978.

Use of the NIH-EPA Chemical Information System in Support of the Toxic Substances Control Act

S. R. HELLER

Environmental Protection Agency, Washington, DC 20460

G. W. MILNE

National Institutes of Health, Bethesda, MD 20014

The Toxic Substances Control Act (TSCA) of 1976, PL-94-469, is an attempt by the Government to provide a general law insuring that chemicals produced in the USA or imported into this country do as little damage and harm to health and the environment as possible.

The NIH/EPA Chemical Information System (CIS), is a collection of computer programs and data bases which has been developed to support scientific and administrative needs of the two Agencies (1). This chapter about the CIS will be primarily devoted to those areas of the CIS which can and are being used by EPA in support of TSCA. The discussions here only relate to the scientific aspects of the system, particularly the area of analytical chemistry.

The quantity of data associated with chemistry has been expanding in the recent decades, but until the recent advent of third generation computers (integrated circuitry computers), handling and using this vast amount of information has been an insuperable problem. With the recent advances in computer technology and electronics, the costs of using computers has come down, while their availability has increased through the use of computer networks, accessible over standard telephone lines. With these features as background, we have been developing a highly interactive, disk-oriented chemical information system of numerical data. This system is readily and inexpensively available to our own Agencies' laboratories, and the public, as well as to our contractors, grantees, and scientific collaborators. With the passage of TSCA, a considerable burden has been placed on industry to report on the manufacture of chemicals. Because it is useful to TSCA, the CIS is being put to broader use that was first envisioned.

The early computer developments in search systems minimized the high cost of mass storage by maintaining data files on magnetic tape rather than drums or disks. Data bases can be stored very inexpensively on tape, but can only be searched sequentially and this is by necessity a slow process because tape is not susceptible to random access. Modern magnetic disks, on the other hand, are random access devices, with considerable storage capacity. The data stored on these devices can be accessed and searched very rapidly. Until recently, the costs of disks, controllers and the other necessary items to use such equipment has precluded their use for large data bases. These costs have decreased markedly in recent years and since the use of disk for data storage permits interactive computing, the Chemical Information System uses disk exclusively for the storage of data.

Interactive computing is a significantly different process from batch, off-line, or even on-line computing and a different philosophy can be used in the design of programs for such work if such a system is to succeed. A major problem that a chemist has in searching a chemical data base, is that the best questions are often not known. An interactive system can provide the answer to a question immediately and this will enable the user to see the deficiencies in the question and to frame a new query. In this way, there can be built a feedback loop in which the scientist acts as a transducer, "tuning" the query until the system reports precisely what is required. The NIH/EPA Chemical Information System, described here in some detail, has been designed around this general approach. The building of a general system requires that all of the known information has been gathered together within one computer system and that this information is all of acceptable accuracy and precision. This is an ideal that the CIS has tried to reach by using the most useful scientific data available in the areas of science that are felt to be most likely to be of value in solving the enormous and diverse problems of improving health and the environment.

System Design

The NIH/EPA CIS consists of a collection of chemical data bases together with a battery of computer programs for interactive searching through these disk-stored data bases. In addition the CIS has a data referral capability as well as a data analysis software system. It can be thought of then, as having four main areas:

1. Numerical Data Bases

2. Data Analysis Software

3. Structure Search System

4. Data Base Referral

The numeric data bases that are part of the CIS include files of mass spectra (2), carbon-13 nuclear magnetic resonance (3), X-ray diffraction data for crystals (4) and powders (5), mammalian acute toxicity data (6), and aquatic toxicity data (7). There are several bibliographic data bases associated directly with the mass spectrometry, X-ray crystallography and nuclear magnetic resonance spectroscopy and have been included within the CIS (8). The analytical programs include a family of statistical analysis and mathematical modelling algorithms (9) and programs for the calculation of isotopic enrichment from mass spectral data (10), second order analysis of nmr spectra (11) and energy minimization of conformational structures (12). Programs that design chemical syntheses are being tested and may, if viable, become part of the CIS in the future (13).

The center of the CIS is the substructure search system (SSS) (14), which allows the user to search a data base of structures (associated with a particular collection of numeric data, such as mass spectra) for occurrences of a particular structure or fragment. With this program, for example, regulatory Agencies considering the problem of collecting data on aromatic bromo-chloro compounds could proceed as follows: The substructure search shown in Figure 1 could be conducted and this would find all occurrences of BrCl compounds in the 25 data bases searched. In turn, by reference to TSCA, ITC, etc. this would lead to information such as the number of chemicals affected, the dollar volume of chemicals affected, and so on. If necessary, a subset of these chemicals could be defined and investigated in further detail.

```
OPTION?  RING
OPTION?  ABRAN  1 AT 1  1 AT 2
OPTION?  SATOM 7
SPECIFY ELEMENT SYMBOL = CL
OPTION?  SATOM 8
SPECIFY ELEMENT SYMBOL = BR
OPTION?  ALTBD  1  2
OPTION?  D
         3 . . 4

          .            .

8BR2                 5

          .            .

        1 . . 6
        ?
        ?
        7CL


OPTION?  EXIM
SPECIFY SEARCH LEVELS TO BE CHANGED
LEVELS = 4
OPTION?  RPROBE
        C??C
     ?        ?
    ?            ?
  C                C??
    ?            ?
     ?        ?
        C??C
          ?
            ?
```

```
                CONDITIONS OF SEARCH
        CHARACTERISTICS TO BE MATCHED          TYPE OF MATCH
TYPE OF RING OR NUCLEUS                         EXACT
NO HETEROATOMS                                 EXACT
SUBSTITUENTS AT   1   2                         IMBED
THIS RING/NUCLEUS OCCURS IN 2715 COMPOUNDS

FILE =   1,  2715 COMPOUNDS CONTAIN THIS RING/NUCLEUS
```

Figure 1. Search for aromatic chloro, bromo compounds in the CIS unified data base

In the area of structure elucidation, if one had evidence that an unknown contained a particular substructure, a search might reveal that there were NMR spectra to compare with such a similar structure, but no IR spectra, suggesting that an NMR spectrum would be more useful than an IR spectrum in attempts to identify the unknown.

As more and more data bases are collected and merged into the SSS, it indirectly becomes a catalog of files/lists/collections containing certain chemicals (15). Recently the structure of the SSS files were reorganized so that this referral capability became much more efficient, using an integrated data base of 25 files as shown in Figure 2.

The entire CIS structure can be viewed, as shown in Figure 3, as "a wheel" of independent numerical data bases, connected or linked together through the SSS "hub", using the Chemical Abstracts Service Registry Number (CAS REGN) as the unique universal chemical identifier for each compound. The use of the CAS REGN to tag all CIS files, was codified in EPA regulation #2800.2 in 1975 (16). With the passage of TSCA in late 1976, the use of the CAS REGN was extended to the TSCA inventory and thus establishes the link between regulatory data and scientific data, both within the CIS and in the literature. In Figure 3, the solid circles represent systems running on commercial systems. The solid box represents a system which is currently being put through its final testing at NIH to "smooth" off any rough edges before it is considered operational and placed in the commercial system. The dotted circles are systems under development. Lastly, the dotted lines to the solid circles refers to operational systems that the CIS can link to, but that are on other computers, either on the same network or on a different network (eg. Tymnet or Telenet).

CIS System Development

A general protocol for updating of CIS components or the addition to the CIS of new components has been established and a schematic diagram of this protocol is shown in Figure 4.

In the first phase, a data base is acquired from one of a variety of sources. Some of the CIS data bases have been developed specifically for the CIS, an example of this being the mass spectral data base (2). Others, such as the Cambridge Crystal File (4), are leased for use in the CIS and still others, such as the X-ray powder diffraction file (5), are operated within the CIS by their owners, in this case the Joint Committee on Powder Diffraction Standards.

In other cases, the information comes from another Government Agency which retains responsibility for the file, its contents and its maintenance. An example of such a file is the NIOSH RTECS.

If the data base is to be made searchable, some reformatting, sorting and inversion of files is usually required and this is carried out on the NIH IBM 360-168, which is well-suited to processing large files of data. Once inverted lists have been prepared, they are transferred to the NIH PDP-10 computer which is primarily a time-sharing computer, and the programs for generating the searchable files and for searching through these files are written. Analytical, data base-independent programs of the CIS are usually written entirely on the PDP-10.

Out of this work, there finally emerges a pilot version of each CIS component. This pilot version is then allowed to run on the NIH PDP-10 and access to it is provided to a small number of people who can log into the NIH computer by tele-

INTEGRATED SSS DATA BASE 3/1/78

FILE	NUMBER OF COMPOUNDS
NIH/EPA—MSSS	25,560
C-13 NMR	3,765
EPA—ACTIVE INGREDIENTS IN PESTICIDES	1,454
PESTICIDES STANDARDS	384
ORD—CHEMICAL PRODUCERS	375
OIL AND HAZARDOUS MATERIALS	858
AEROS/SAROAD	65
AEROS/SOTDAT	572
STORET	234
CHEMICAL SPILLS	577
TSCA INVENTORY CANDIDATE LIST	33,579
NIMH—PSYCHOTROPIC DRUGS	1,686
SRI-PHS LIST 149 OF CARCINOGENS	4,448
NBS—SINGLE CRYSTAL FILE	18,362
HEATS OF FORMATION OF GASEOUS IONS	3,169
GAS-PHASE PROTON AFFINITIES	454
NSF—RANN POLLUTANT FILE	225
FDA—PESTICIDE REFERENCE STANDARDS	613
CPSC—CHEMRIC MONOGRAPHS	1,000
CAMBRIDGE UNIVERSITY CRYSTAL DATA	10,018
EROICA THERMODYNAMIC DATA	4,492
MERCK INDEX	8,894
ITC—INTERNATIONAL TRADE COMMISSION	9,194
NIOSH—REGISTRY OF TOXIC EFFECTS OF CHEMICAL SUBSTANCES	19,908
NFPA—HAZARDOUS CHEMICALS	397

Figure 2. List of the current 25 collections which comprise the CIS unified data base

CIS COMPONENTS 3/1/78

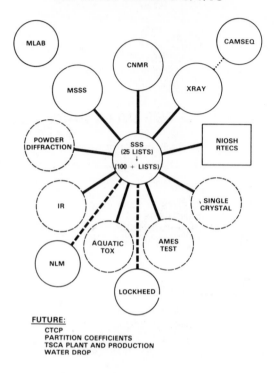

Figure 3.   The structure of the CIS with the CAS registry number linking

ADDITION OF A COMPONENT TO THE CIS

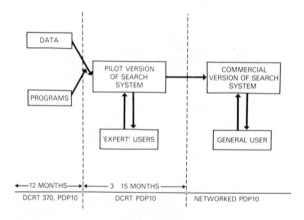

Figure 4.   Protocol for adding a component to the CIS

phone, using long distance calls if necessary. These users are provided with free computation and in return, they test the component thoroughly for errors and deficiencies. Such problems are reported to the development team, which attempts to deal with them. Depending upon the size and complexity of the component, this testing phase can last as long as eighteen months.

When testing is complete, the entire component is exported to a networked PDP-10 in the private sector and the version on the NIH computer is no longer maintained. The component in the private sector is available to the general scientific community, including Government Agencies, and is used on a fee-for-service basis. In this phase, the Government retains no financial interest in the component; it is 'managed' by a sponsor outside the U.S. Government. The Netherlands Information Combine in Holland, for example, maintains the Carbon-13 Nuclear Magnetic Resonance (CNMR) search system on the network. Advice and consultation between such sponsors and NIH/EPA personnel continues, but the U.S. Government does not subsidize the routine operation of CIS components in the private sector. In fact, various Government agencies of the Government are actually users of the CIS and they pay, like any other user, according to their use of the system. Charges for use of these components must be designed to cover costs, and if the component attracts insufficient use at these prices, then it may not be viable and its sponsor is free to discontinue its support.

Computer Facilities Used by the CIS

Programs of the CIS have usually been designed for use with a DEC PDP-10 computer system. The reason for this is that the PDP-10 is one of the better time-sharing systems available and has been adopted by a number of commercial computer network companies as the main vehicle for their networks. Transfer of a program from the NIH PDP-10 to a network PDP-10 is usually fairly straightforward, and use of a networked computer is favored because the alternative philosophy of exporting programs and data bases to locally operated PDP-10 computers is less workable and contains a number of deficiencies that are overcome by a network. Most important among these is the fact that use of a networked machine means that data bases need only be stored once, at the center of the network. A great deal of money is thus saved because duplicate storage is not necessary. Further, a single copy of a data base is easy to maintain, whereas updating of a data base that resides on many computers is virtually impossible. Finally, communications between systems personnel and users is very simple in a network environment, as is monitoring of system performance.

For these and other reasons, the policy of disseminating the CIS via networked PDP-10 computers was adopted at the outset and has proved to be quite successful. A typical American network of this sort has something under 100 nodes — i.e. local telephone call access is available in about 100 locations. These are mainly in the U.S., but a substantial number will be found in Europe. Further, some computer networks are now themselves interfaced to the Telex network, thus making their computer systems available worldwide. Irrespective of one's location, the cost of access is somewhere between $7 and $15 per hour, depending upon the transmission speed used and also on the time of day. Networks typically offer 110, 300 and 1200 baud service and the response time of the system is usually negligibly small.

The only equipment that is required to establish access to a computer network, is a telephone-coupled computer terminal. Typewriter terminals are becoming very common and are also becoming relatively inexpensive. Such a terminal can be purchased from a variety of manufacturers for between $1,000 and $3,000 and in general, will operate at 300 baud (30 characters/second). A cathode ray terminal, capable of running at 1200 baud can be purchased for as little as $2,000. Any equipment of this sort can usually be leased or purchased.

Components of the CIS

A. Mass Spectral Search System (MSSS). The Mass Spectral Search System (MSSS) is the oldest component of the CIS. The first version of MSSS was developed in 1971, and the system has been used as a prototype for more recently designed components. Developed as a joint effort between NIH, EPA, NBS (National Bureau of Standards) and the Mass Spectrometry Data Centre (MSDC) in England, the current MSSS data base contains about 30,000 mass spectra representing the same number of compounds. This has been derived from an archival file containing some 60,000 spectra of the same 30,000 compounds (17). Computer techniques have been employed to assign every spectrum a quality index (18) and where duplicate spectra appear in the archive file, only the best spectrum is used in the working file. All compounds in the archive have been assigned Chemical Abstracts Service (CAS) registry number, a unique identifier that is used to locate duplicate entries for the compounds, find the compound in other CIS files and provide structure and synonym lookup capabilities throughout the CIS.

Searches through the MSSS data base can be carried out in a number of ways. With the mass spectrum of an unknown in hand, the search can be conducted interactively, as is shown in Figure 5. In this search the user finds that 24 data base spectra have a base peak (minimum intensity 100% maximum intensity 100%) and an m/e value of 344. When this subset is examined for spectra containing a peak at m/e 326 with intensity of less than 10%, only 2 spectra are found. If necessary, the search can be continued in this way until a manageable number of spectra are retrieved as fulfilling all the criteria that the user cited. These answers can then be listed as is shown. Alternatively, the data base can be examined for all occurrences of a specific molecular weight or a partial or complete molecular formula. Combinations of these properties can also be used in searches. Thus all compounds containing for example, five chlorines and whose mass spectra have a base peak at a particular m/e value can be identified.

In contrast to these interactive searches, which are of little appeal to those with large numbers of searches to carry out, there are available two batch-type searches which accept the complete spectrum of the unknown and examine all spectra in the file sequentially to find the best fits. These are the Biemann and PBM search algorithms. Spectra can be entered from a teletype, but in a more common arrangement, a user's data system can be connected to the network for this purpose and the unknown spectra can be down-loaded into the network computer for use in the search, which can be carried out at once, or, if preferred, overnight at 30% of the cost.

Once an identification has been made, the name and registry number of the data base compound are reported to the user. If necessary, the data base spectrum can be listed or, if a CRT terminal is being used, plotted, to facilitate direct comparison of the unknown and standard spectra.

The MSSS has been widely available through computer networks since 1971 and is currently resident upon the ADP-Cyphernetics network where, every month, over 3000 searches and 2,000 other transactions, such as retrievals, are carried out by approximately 325 laboratories. All searches in the MSSS are transaction priced at be-

```
USER RESPONSE:PEAK

TYPE PEAK,MIN INT,MAX INT
CR TO EXIT,' 1 FOR ID#,REGN,QI,MW,MF AND NAME

USER:344,100,100
    # REFS   M/E PEAKS

        24       344

NEXT REQUEST: 326,0,10
    # REFS   M/E PEAKS

         2       344 326

NEXT REQUEST: 1
    ID#           REGN  QI MW   MF                          NAME

  23455       19594913 624 344 C22H3203              A-Nor-5.alpha.-
                                             androstan-3-one, 17.beta.-hydr
                                             oxy-5-vinyl-, acetate (8CI)
  25330       18326153 974 386 C24H3404             Podocarpa-8,11,
                                             13-triene-3.beta.,12-diol, 13-
                                             isopropyl-, diacetate (8CI)
```

Figure 5.   PEAK search in the MSSS

tween $1 and $10 and in addition to these charges and the connect time charge, users must each pay an annual subscription fee of $300. This fee is used to defray the annual disk storage charges which are paid in advance by the sponsor of the MSSS, the Department of Industry of the British Government.

B. Mass Spectrometry Literature (BULL) Search System. The accumulated files of the Mass Spectrometry Bulletin, a serial publication of the Mass Spectrometry Data Centre, Aldermaston, England, have been made the basis of an on-line search system (27). The Bulletin, which since 1967, has collected about 60,000 citations to papers on mass spectrometry, may be searched interactively for all papers by given authors, all papers dealing with one or more specific subjects or with one or more particular elements. In addition, citations dealing with general index terms may also be retrieved. Simple Boolean logic is available, and thus searches may be conducted for papers by Smith and Jones, or Smith but not Jones, and so on. Citations retrieved may be limited to specified publication years, between 1967 and the present. The interactive nature of the search provides great control to the user. One can learn within a few minutes that while there are in the Bulletin, 463 papers dealing with mass measurement for example, and 678 on chemical ionization, only 8 report on mass measurement in chemical ionization mass spectra. Similarly, one can rapidly discover that although there are 532 papers dealing with carbon dioxide, only 1 of these was presented at the 1975 NATO meeting in Biarritz.

No numerical codes are used by the system. A search for a specific subject can be carried out by entering the subject word itself. If the word 'mass' is entered, searches for 7 terms (all those containing the fragment 'mas' i.e. mass spectra, mass discrimination, mass measurement, etc.) are conducted and the user is asked to select the one of interest. In this way knowledge of the correct subject words or of their correct spelling is not necessary.

The Bulletin search system is a component of the MSSS, described above, and as such, is accessible as indicated in the previous section on the MSSS.

C. Carbon-13 Nuclear Magnetic Resonance (CNMR) Spectral Search System. The data base that is used in the CNMR search system consists currently of 4,100 CNMR spectra (3). As in the case of the MSSS, every compound has a CAS registry number, and all exact duplicate spectra have been removed from the file. A specific compound may still appear in this file more than once, however, because its CNMR spectrum may have been recorded in different solvents. The CNMR file is still small but is growing at a fairly steady rate and should benefit considerably from recent international agreements to the effect that all major compilations of CNMR data will, in the future, be pooled.

Searching through this data base, as in the case of the MSSS, can be interactive or not. In the interactive search, a user enters a shift, with an acceptable deviation, and the single frequency off-resonance decoupled multiplicity, if that is known. The algorithm reports the number of file spectra fitting one or both of these criteria. The names of the compounds whose spectra have been retrieved can be listed, or alternatively, the list can be reduced by the entry of a second chamical shift. A search for spectra of compounds having a specific molecular formula can also be carried out, but there is no capability for searching on molecular weight, a parameter of little relevance in CNMR spectroscopy.

If an interactive search is not appropriate to the problem at hand, a batch type of search through the data base is available using the techniques described by Clerc et al. (19). To institute such a search, the user enters the all the chemical shifts from the unknown and starts the search. The entire unknown spectrum is compared to every entry in the file and the best fits are noted and reported to the user. This program searches for the absence of peaks in a given region as well as for the presence of peaks and thus has the capability of finding those compounds which are structurally similar to the material that gave the unknown spectrum.

When a search is completed, the user is provided with the accession numbers of spectra that fit the input data. The names and CAS registry numbers of the compounds in question will also be given. If more information is required, the complete entry for a given accession number can be retrieved. This includes a numbered structural formula, the name, molecular formula and registry number of the compound, experimental data pertaining to the spectrum and the entire spectrum, together with single frequency off-resonance decoupled multiplicities and (if available, which is about 60% of the time) relative line intensities and assignments.

This CNMR search system has recently been made available on the ADP-Cyphernetics network. Searches are all transaction priced at $1 — $3, and the subscription fee for the system is $100 per year.

D. X-ray Crystallographic (CRYST) Search System. This is a series of search programs working against the Cambridge Crystal File (4), a data base of some 15,000 compounds for which full atomic co-ordinate data available, and over 27,000 bibliographic entries dealing with published crystallographic data mainly for organic compounds. The entry for each compound contains the compound name, its molecular weight and CAS Registry number, the space group in which it crystallizes and the parameters of the unit cell of the crystal. The file may be searched on the basis of any of these parameters as shown in Figure 6, which shows a search for any compounds that crystallize in space group P 1 and have molecular weights between 250 and 300. As can be seen, there are 98 entries with the correct space group (temporary file 1) and 867 with molecular weight between 250 and 300 (temporary file 3). The intersection of these files reveals that only 3 compounds (temporary file 4) meet both specifications, and the first of these compounds, crystal sequence number 4413, is listed.

All the compounds in this file have been registered (at the time Figure 6 was prepared the CAS data was not yet merged into the CRYST system) by the CAS and their connection tables have been merged into the file. This data base is therefore searchable on a structural or substructural basis, as are all the other files of the CIS.

Once an entry of interest in the Cambridge X-ray file has been located by one of the search programs, its 'crystal sequence number' can be used to retrieve the appropriate literature reference, structure, or co-ordinate data or both.

This file is available for general use via the ADP-Cyphernetics network. Currently, the charging of options in this system is not transaction-priced. Enough statistics are now available to indicate that all searches other than structural searches cost less than $2.00 and that the structural searches cost possibly from $2 to $10.

E. X-ray Crystal Data Search System. The National Bureau of Standards (NBS) has collected a file of data pertaining to some 24,000 crystals, including those in the Cambridge file described above (20). The data in the NBS file include the cell parameters, the number of molecules, Z, in the unit cell, the measured and calculated densities of the crystal and two determinative ratios, such as A/B and A/C. Every com-

```
*SPGR
SPACE  GROUP  SYMBOL
>  P  1
FILE  =  1  REFERENCES  =      98  ITEM  =  P  1
*SMOLS  1
TYPE  MOLECULE  WEIGHT  RANGE

>250,300
FILE  =  3      MERGED  REFERENCES  =  867

*INTER  1  3
FILE  =  4        INTERSECTED  REFERENCES  =        3
SOURCE  FILES  WERE:      1      3
*SSHOW  4
START  WITH  NTH  REFERENCE    (1)=1
SHOW EVERY  NTH  REFERENCE    (1) =1
STRUCTURE        1  CRYSTAL  SEQUENCE        4413

I              C                        C * * N * * C
                *                        *        *  *
                  *                    *          *    *
                    C * * O * * C        *        C
                  *                    *          *    *
                *                        *        *  *
              O                        C * * C
                                        *      *
                                        *      *
                                        C * * C
```

Figure 6.   Space group and molecular weight search in the Cambridge crystal
data base

pound in the file is identified by its name, molecular formula and registry number, and the file can be structurally searched by the CIS substructure search system as is described below.

Searches through this data base for crystals with specific space groups, or densities are being developed and it will be possible to locate crystals with unit cells of given dimensions. It is hoped that this may prove to be a very rapid method of identifying compounds from the readily measured crystal properties.

F. X-ray Powder Diffraction Search System. Compounds that fail to crystallize may still be examined by X-ray diffraction, because non-crystalline materials, as powders, give characteristic diffraction patterns. A collection of powder diffraction patterns proves to be a very effective means by which to identify materials and indeed, one of the very earliest search systems in chemical analysis was based upon such data by Hanawalt (21) over forty years ago. The importance of these data in TSCA can be seen by examining the TSCA Inventory regulations for treatment of confidential chemicals (22). Section 710.7 of these regulations indicates that EPA intends to rely on powder diffraction data to assure the validity and seriousness of a manufacturers request for treating information on a chemical as confidential.

The data base of some 27,000 powder diffraction patterns that is used in the CIS (5) is in fact a direct descendant of that with which Hanawalt carried out his pioneering work. A problem that arises in connection with this particular component stems from the fact that powders, as opposed to crystals, are frequently impure and so the patterns that are obtained experimentally are often combinations of one or more file entries. A reverse searching program, that examines the experimental data to see if each entry from the file is contained in it, has been written after the general approach of Abramson (23), and seems to cope with this particular difficulty. It is currently running in test on the NIH PDP-10 and will be made available to the scientific community during the latter part of 1978.

G. NIOSH RTECS Search System. The National Institute for Occupational Safety and Health (NIOSH), created in 1970, is required by law to prepare a list containing all the toxic effects of chemicals that can be found to have been recorded (24). The Registry of Toxic Effects of Chemical Substances (RTECS) is the data base created and updated annually by NIOSH to fulfill the provisions of this law. In 1977 the data base consisted of some 25,000 chemicals and the toxicity associated with these chemicals.

The NIOSH RTECS is the first non-spectroscopic CIS data base and has proven to be a very valuable addition to the CIS. Interest in the data base has been shown by many groups within EPA involved in the implementation of TSCA. For example, work is now underway to link spectral data with the NIOSH toxicity data so that as a result of a mass spectral identification, the EPA lab can quickly be informed if the chemical identified is toxic and hence requires immediate action.

The RTECS data base can be searched in a number of ways, including NIOSH number, CAS Registry number, type of animal tested, route of dosage, LD50, LC50, etc. The NIOSH RTECS file is also linked to the SSS so that structure-activity correlation work can be performed.

An example of a NIOSH RTECS search is shown in Figure 7. In this example, a search is being performed for all oral rodent LD50 toxicity data with values less than 75 ug./kg. of which the system indicates there are 3. The bottom of Figure 7 shows these three references, with the NIOSH number, toxicity data and the literature citation for that measurement.

```
OPTION?SEARCH
ANIMAL TYPE?ROD
DOSAGE METHOD?ORL
TYPE OF MEASURE?LD50
DO YOU WANT ALL VALUES OF TOXICITY?NO
UPPER NUMERIC LIMIT AND UNITS?1 MG/KG
LOWER NUMERIC LIMIT AND UNITS?0 MG/KG
SEARCH OF THE RTECS DATA BASE FOR:
ANIMAL =  ROD    DOSAGE METHOD =   ORL
MEASURE =  LD50 AND TOXIC EFFECT =   ALL
BETWEEN THE LIMITS    1       MG/KG AND    0       MG/KG
YIELDS      41 COMPOUNDS ON FILE      1
OPTION?DSHOW 1
DISPLAY HOW MANY? (TYPE E TO EXIT)3
```

```
CAS NUMBER =        51183 NIOSH NUMBER = XZ21000
    ORL-RAT LD50:     1 MG/K    TFX:           BWHOA6      31,721,65
    IPR-RAT LD50:     1 MG/K    TFX:           JPETAB      100,398,50
    IPR-RAT TDLO:   500 UG/K    TFX:TER        CPCHAO      18,307,62
    SCU-RAT TDLO:    10 MG/K    TFX:NEO        ANYAA9      68,750,58
    IMS-RAT LD50:  1500 UG/K    TFX:           CLDND*
    ORL-MUS LD50:    15 MG/K    TFX:           JPETAB      100,398,50
    SKN-MUS TDLO:    10 MG/K    TFX:NEO        BJCAAI      9,177,55
    IPR-MUS LD50:  2800 UG/K    TFX:           JPETAB      100,398,50
    IPR-MUS TDLO:   200 UG/K    TFX:MUT        NATUAS      219,385,68
    IPR-MUS TDLO:     3 MG/K    TFX:NEO        BJPCBM      6,357,51
    SCU-MUS TDLO:  1350 UG/K    TFX:TER        JEEMAF      11,689,50
    IMS-MUS LD50:  1500 UG/K    TFX:           CLDND*
    ORL-DOG LDLO:     1 MG/K    TFX:           JPETAB      100,398,50
    IVN-DOG LDLO:   400 UG/K    TFX:           JPETAB      100,398,50
    IVN-MKY LDLO:   100 UG/K    TFX:           CCSUBJ      2,202,65
    IVN-CAT LDLO:     1 MG/K    TFX:           JPETAB      100,398,50
    1,3,5-Triazine, 2,4,6-tris(1-aziridinyl)- (9CI)
    C9H12N6
```

```
CAS NUMBER =        80126 NIOSH NUMBER = JO63000
    ORL-RAT LD50:   111 UG/K    TFX:           PCOC**      -,1122,-
    ORL-MUS LDLO:   200 UG/K    TFX:           12VXA5      8,1028,68
    UNK-MAM LD50:   100 UG/K    TFX:           30ZDA9      -,263,71
    2,6-Dithia-1,3,5,7-tetraazatricyclo[3.3.1.13,7]decane, 2,2,6
    6-tetraoxide (9CI)
    C4H8N4O4S2
```

```
CAS NUMBER =        98055 NIOSH NUMBER = CY31500
    ORL-RAT LDLO:    50 MG/K    TFX:           JPETAB      93,287,48
    ORL-MUS LD50:   270 UG/K    TFX:           CLDND*
    IVN-RBT LD50:    16 MG/K    TFX:           JPETAB      80,93,44
    Arsonic acid, phenyl- (9CI)
    C6G7AsO3
```

Figure 7.   Search for acute toxicity data

H. Substructure Search System (SSS). All the compounds in the files of the CIS have been assigned a registry number by the CAS. The registry number is a unique identifier for that compound, and may be used to retrieve from the CAS Master Registry of over 4,000,000 entries, all the synonyms that the CAS has identified for the compound, these being names that have been used for the compound in addition to the name used in the CAS 9th Collective Index. Further, the registry number can be used to locate in the CAS files, the connection table for the compound's structure. This is a two-dimensional record of all atoms in the molecule together with the atoms to which each is bonded and the nature of the bonds (25). This connection table is the basis of the substructure search component of the CIS (25).

The purpose of the SSS is to permit a search for a user-defined structure or substructure through data bases of the CIS. If a substructure is found to be in a CIS data base, then, armed with its CAS Registry number, the user can access that file and locate the compound and hence retrieve whatever data is available for it.

There are a number of ways to search the CIS Unified Data Base. The main ones are:

1. Name/Fragment Name Search (NPROBE)

2. Nucleus/Ring Search (RPROBE)

3. Fragment Search (FPROBE)

4. Structure Code (CIDS) Search (SPROBE)

5. Molecular Weight, Molecular Formula, Partial Formula

6. Total Atom-by-atom, Bond-by-bond search (SSS)

7. Total or Full Structure Search (IDENT)

Before going into the main features of the SSS, one new feature is worth mentioning. That is the name or nomenclature search (NPROBE). While structure searching is very important and cannot be replaced by other methods (such as fragment searching, linear notations or name searching), being able to search for a chemical by name or partial name, is quite useful in many cases. In particular, if one wishes to search for a drug or pesticide, all of which have simple and short common or trade names, a name search is very useful. The main reason is that drugs and pesticides are often complex polycyclic structures, difficult to draw. In the example shown in Figure 8, a name search is conducted for the drug Dimedrol. The program then is commanded (using the SSHOW command) to print out the files in which this one chemical containing the name fragment Dimedrol appears, along with its molecular formula, structural diagram and correct Chemical Abstracts Index name, as well as the synonyms associated with the chemical.

As the first step in a structure search, the user must define the substructure of interest to the computer. This is done with a family of structure generation programs which can for example, create a ring of a given size, a chain of a given length, a fused ring system and so on. Branches, bonds and atoms can be added and the nature of bonds and atoms can be specified. In the absence of a definition, an atom is presumed to be carbon. As the query structure is developed using these commands, the computer stores the growing connection table. If the user wishes to view the current structure at any point, the display command (D) can be invoked. This command, using the current connection table, generates a structure diagram similar to those in Figures 9, 10 and 11.

```
OPTION? NPROB
FULL NAME ONLY(Y/N)?
Y
SPECIFY NAME (CR TO EXIT):
DIMEDROL
FILE   5,  1 COMPOUNDS HAVING NAME/FRAGMENT
DIMEDROL
SPECIFY NAME (CR TO EXIT):

OPTION? SSHOW 5
STRUCTURE   1 CAS REGISTRY NUMBER 147-24-0
TSCA CANDIDATE LIST: R062-2752
U.S. INTERNATIONAL TRADE COMMISSION
PHS-149 CARCINOGENS: B0276
NIOSH RTECS: KR70000
```

C17H21NO.CLH

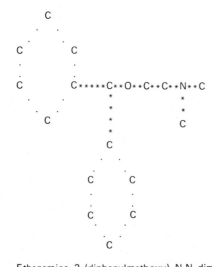

```
Ethanamine, 2-(diphenylmethoxy)-N,N-dimethyl-, hydrochloride (9CI)
Ethylamine, 2-(diphenylmethoxy)-N,N-dimethyl-, hydrochloride (8CI)
.beta.-Dimethylaminoethyl benzhydryl ether hydrochloride
Allergan
Allergival
Bena
Benadril hydrochloride
Benadryl
Benzhydramine hydrochloride
Denydryl
Difenhydramine hydrochloride
Dimedrol
_____
OPTION?
```

Figure 8.   Name search (NPROBE) for dimedrol

This can be printed at a conventional terminal.

When the appropriate query structure has been generated, a number of search options can be invoked to find occurrences of this query structure in the data base. The two most useful search options are the fragment probe and the ring probe. The fragment probe will search through the assembled connection tables of the data base for all occurrences of a particular atom-centered fragment, i.e. a specific atom, together with all its neighbors and bonds. The user may specify particular fragments which are thought to be fairly unique and characteristic of the query structure. Alternatively, a search for every fragment in the query structure may be requested. A fragment probe is shown in Figure 9. The query structure contains only one relatively unique node, C3, and this is the one which is sought in the data base. It is found to occur 980 times and a temporary file of just those particular entries is stored as file #2. This can be accessed by the user either for the purpose of listing its contents, as is shown in the figure, or to intersect it with other scratch files.

The ring probe search is a search for all structures in the data base containing the same ring or rings as the query structure. A ring that is considered to be an answer to such a query must be the same size as that in the query structure. It must also contain at least as many heteroatoms number of non-carbon atoms (heteroatoms) but the nature of the heteroatoms can be required by the user to be the same or different to that in the query structure. Bonding is not considered in an RPROBE search. Thus with a query structure of pyrrole, the only 'exact' answer is pyrrole but the user may permit the retrieval of 'imbedded' answers which would include furan, tetrohydrofuran and thiophene. An example of a ring probe search is given in Figure 10. Here the query structure is a 3,4-dichlorofuran, but imbedded matches for heteroatoms and substituents have been allowed and so the list of 304 answers will include any disubstituted pyrrole as well as any disubstituted furan and so on. Higher substitution will also be permitted.

In addition to these structural searches, there are a number of 'special properties' searches that often prove to be very useful as a means of reducing a large list of answers resulting from structure searches. The special properties searches include searches for a specific molecular weight or range of molecular weights and a search for compounds containing a given number of rings of a given size. Searches may also be conducted for the molecular formula corresponding to the query structure, or for a different, user-defined molecular formula. This may be specified completely or partially and the number of atoms of any element may be entered exactly or as a permissible range.

If one's purpose is to determine only the presence or absence in a data base of a specific structure, this can be accomplished with the search option 'IDENT', as is shown in Figure 11. This program hash-encodes the query structure connection table and searches through a file of hash-encoded connection table for an exact match. The search, which is very fast by substructure search standards, has been designed specifically for those users who, to comply with the Toxic Substances Control Act [26], have to determine the presence or absence of specific compounds in Environmental Protection Agency files.

Finally, if one has completed ring probe and fragment probe searches for a specific query structure and is still confronted with a sizeable file of compounds that satisfy the criteria that were nominated, a sub-structure search through this file may be carried out. This involves an atom-by-atom, bond-by-bond comparison of every struc-

OPTION? FPROBE 2

TYPE E TO EXIT FROM ALL SEARCHES,
T TO PROCEED TO NEXT FRAGMENT SEARCH

FRAGMENT:

           8BR????2C.....1C
                    .
                    .
                    .
                   3C

REQUIRED OCCURRENCES FOR HIT :   1
THIS FRAGMENT OCCURS IN   229 COMPOUNDS

FILE = 3,   229 COMPOUNDS CONTAIN THIS FRAGMENT

Figure 9.   SSS fragment probe search

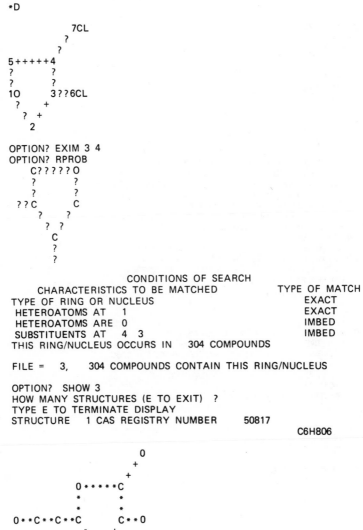

```
*D

                7CL
             ?
           ?
5 + + + + + 4
?         ?
?         ?
10       3 ? ? 6CL
  ?    +
    ? +
    2


OPTION? EXIM 3 4
OPTION? RPROB
    C ? ? ? ? ? O
    ?         ?
    ?         ?
  ? ? C       C
      ?     ?
      ? ?
        C
        ?
        ?
```

CONDITIONS OF SEARCH

CHARACTERISTICS TO BE MATCHED	TYPE OF MATCH
TYPE OF RING OR NUCLEUS	EXACT
HETEROATOMS AT 1	EXACT
HETEROATOMS ARE 0	IMBED
SUBSTITUENTS AT 4 3	IMBED

THIS RING/NUCLEUS OCCURS IN   304 COMPOUNDS

FILE =   3,   304 COMPOUNDS CONTAIN THIS RING/NUCLEUS

```
OPTION? SHOW 3
HOW MANY STRUCTURES (E TO EXIT)  ?
TYPE E TO TERMINATE DISPLAY
STRUCTURE   1 CAS REGISTRY NUMBER      50817
                                              C6H8O6
```

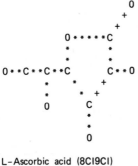

L-Ascorbic acid (8Cl9Cl)

Figure 10.   SSS ring/nucleus probe search

```
OPTION?  D

    10CL  7O
    ?     +
    ?     +
8CL3??1??2P?5O?11
    ?  ?  ?
    ?  ?  ?
    9CL4O 6O
        ?
        ?
        12
```

OPTION? IDENT

TOTAL PROTON COUNT FOR THIS STRUCTURE IS
        (P FOR PROGRAM ESTIMATE) : P
TOTAL PROTON COUNT BASED UPON NORMAL CONDITIONS IS 8
ARE THERE ANY ABNORMAL VALENCE OR CHARGE CONDITIONS WHICH
        WOULD AFFECT THIS COUNT (Y/N) ? N
PROTON COUNT FOR NODE 2 (D TO DISPLAY STRUCTURE) ?
FILE 10, THIS STRUCTURE IS CONTAINED IN   1 COMPOUNDS.
OPTION? SSHOW 10
STRUCTURE   1 CAS REGISTRY NUMBER 52-68-6
TSCA CANDIDATE LIST: R001-5032
EPA PESTICIDES – ACTIVE INGREDIENTS: 57901
EPA OHM/TADS: 72T16519

CAMBRIDGE XRAY CRYSTAL: 52-68-6.01
MERCK INDEX
EPA PESTICIDES – ANALYTICAL REF. STNDS.: 6780
EPA CHEMICAL SPILLS
FDA/EPA PESTICIDES REF. STANDARDS: 48
PHS-149 CARCINOGENS: C0147
NIOSH RTECS: TA07000

                                                        C4H8C13O4P

    CL   O
    *   +
    *   +
CL*C**C**P**O**C
    *   *   *
    *   *   *
    CL O   O
            *
            *
            C
```

Phosphonic acid, (2, 2, 2-trichloro-1-hydroxyethyl)–, dimethyl ester (8C
        I9CI)
Agroforotox
Anthon
Bayer L 13/59
Chlorofos

*Figure 11.   Complete structure (IDENT) search*
```

ture and will retrieve any compound which contains the query structure imbedded in it.

The substructure search system is currently operating on a unified data base of 25 files which are given in Figure 2. The SSS data bases are in the process of being updated with an addition 29 files, which are shown in Figure 12. This update is scheduled for the spring of 1978. The whole system is available for general use on the Tymshare computer network. A subscription fee of $150 per year must be paid for use of the system and the only other charges are the connect time charge and the searching costs which range upwards from between $3.00 — 5.00 for an identity search.

I. X-Ray Literature Search System. The data base used in the X-ray crystallographic search system described in (c) above possesses complete literature references to all entries in the file [4]. This information has been made the basis of a system for searching by author, title word, etc, the literature pertaining to the X-ray diffraction study of organic molecules.

As in the Mass Spectrometry Bulletin Search System, it is possible to search for papers by a specific author or authors, and papers that appeared in given years in given journals may also be retrieved. Additionally, papers may be located on the basis of specific words appearing in their titles. These words may be truncated by the user and so the fragment 'ERO' will retrieve papers with the word 'STEROID' in their titles or papers whose titles use the word 'MEROQUININE'. The system generates scratch files from searches, as in the substructure search system, and files can be intersected upon request with 'AND' or 'NOT' operators. Thus one can, for example, retrieve all papers published in Acta Crystallographica since 1970 by Atkins, excluding specifically those on corticosteroids.

Once a paper of interest has been identified, all the crystallographic information in that paper can be examined because the crystal serial number of the paper can be used in the crystallographic search system to retrieve that information. Alternatively, the CAS Registry number of any particular compound can be used to retrieve any data of interest on that compound from other files of the CIS.

J. Proton Affinity Retrieval Program. With the current high level of interest in chemical ionization mass spectrometry, there is a need for a reliable file of gas phase proton affinities. No data base of this sort has previously been assembled and for these reasons, the task of gathering and evaluating all published gas phase proton affinities has been undertaken by Rosenstock and co-workers at NBS. This file [28], which has about 400 critically evaluated gas phase proton affinities drawn from the open literature, and can be searched on the basis of compound type or the proton affinity value. It will be appended to the MSSS and the bibliographic component will be merged with the Mass Spectrometry Bulletin Search System.

K. NMR Graphical Interactive Spectrum Analysis (GINA) Program. Many proton nmr spectra can be satisfactorily analyzed by hand, and such first order analysis is, in these cases, a quite satisfactory way of assigning chemical shifts and coupling constants to the various nuclei involved. In certain cases however, so-called second order effects become important and as a result, more or fewer spectral lines than are indicated by first order considerations will result. A way to analyse such spectra is to estimate the various coupling constants and chemical shifts and then, using any of a variety of standard computer programs [11], calculate the theoretical spectrum corresponding to these values. The calculated spectrum can be compared to the observed spectrum and a new estimate of the data can be made. In this way, by a series of

29 New Files for
the NIH/EPA CIS UDB

U.S. Coastguard Chemical
Properties File.

EPA IERL Non-Criteria Pollutant
Em·;sions.

EPA, Section 111A of the Clean
Air Act.

EPA, Office of Air Quality,
Permissible Standards,
Criteria Pollutants.

EPA, Office of Water Supply,
File of Drinking Water
Pollutants.

EPA, Pollutant Strategies Branch,
Selected Organic Air
Pollutants.

EPA, Effluent Guidelines Consent
Decree List

EPA, Section 112 of the Clean Air
Act.

EPA, ORD, Gulf Breeze, List of
Chemicals.

EPA, Carcinogen Assessment Group
List of Chemicals.

EPA, RPAR Candidates Chemical
Review Schedule List.

EPA, OTS Status Assessments.

EPA, Standing Air Monitoring
Work Group List of Non—
Criteria Pollutants.

EPA, ORD—OHEE Laboratory
Chemicals.

EPA, List of Potentially
Hazardous Chemicals from Coal
and Oil.

California OSHA List of Chemical
Contaminants.

WHO, Food and Agriculture
Organization, List of Pesticides.

EPA, IERL, Organic Chemicals
in Air.

NCI, Public List of Known
Carcinogens.

NCTR, Potential Industrial
Carcinogens and Mutagens.

EPA, IERL, List of Environmental
Carcinogens.

EPA, OPP, Pesticide Literature
Searches.

NIEHS, Laboratory Chemicals.

Toxic and Hazardous Industrial
Chemicals Safety Manual.
International Technical
Information Institute, Tokyo.

List of Teratogenic Chemicals.
Medical Information Center,
Karolinska Institute, Stockholm.

EPA, List of Hazardous Pesticides.

EPA, Mutagenicity Studies.

CITT, List of Candidates.

EPA, TSCA Section 8e, List of
Chemicals.

Figure 12. Collections/files being added to the CIS unified data base

successive approximations, the correct coupling constants and chemical shifts can be determined.

The CIS component GINA (Graphical Interactive Nmr Analysis) which is based upon the programs developed by Johannesen et al [29], permits these operations in real time in an interactive fashion. The program is designed for use with a vector cathode ray tube terminal upon which each new theoretical spectrum can be displayed for comparison by the user with the observed spectrum. The program has been available at NIH for over four years and is currently being exported to a computer network in the private sector. The cost of using the program is not yet well established because it is subject to wide variations.

L. Mathematical Modelling System (MLAB). MLAB is a program set, developed by Knott at NIH [9], which can assimilate a file of experimental data, such as a titration curve, for example, and perform on it any of a wide variety of mathematical operations. Included amongst these are differential and integral calculus, statistical analysis (mean and standard deviation, curve and distribution fitting and linear and non-linear regression analysis). Output data can be presented in any form, but the PDP-10-resident program is especially powerful in the area of graphical output. Data can be displayed in the form of two-or three-dimensional plots which can be viewed and modified on a CRT terminal prior to pen-and-ink plotting.

M. Isotopic Label Incorporation Determination (LABDET). Radioisotopes are particularly well-suited to labelling studies because they can be very easily detected at very low levels. In recent years however, there has been increasing concern about the shortcomings of radioisotopes in medical research. Current standards in fact, take the position that the use of radioisotopes such as carbon-14 in children and women of child-bearing age is precluded. Consequently, it is not possible to study the metabolism of drugs in such patients using radioisotopes, and this leads to some difficulty because it is only in such patient groups that the metabolism of drugs such as oral contraceptives is of relevance.

Much effort has gone into studies of the possibilities of carbon-13 as a surrogate for carbon-14, and this type of work applies also to problems involving oxygen and nitrogen which have stable isotopes, but no convenient radioisotopes. Mass spectrometry is the best general method for detection and quantitation of stable isotopes in molecules but there is a serious problem involved in its application. This is that stable isotopes such as carbon-13, nitrogen-15 and oxygen-18 occur naturally as minor components of the natural elements. This is most pronounced in the case of carbon. Naturally-occurring carbon is about 99% C-12 but a small and variable amount of all natural carbon is C-13. This creates a 'background' against which determinations of isotope levels in labelled compounds must be measured. The purpose of LABDET [10] is to compare the mass spectrum of an unlabelled compound with that of the same compound isotopically enriched. This is usually done using the molecular ion region of the spectra. The program calculates an estimate of the level of incorporation of isotope and then calculates a theoretical spectrum which can be compared with the actual spectrum. The theoretical spectrum is then adjusted and a further comparison is made and in this way, the program proceeds through a predetermined number of iterations, finally calculating the correlation coefficient between the observed spectrum and the best theoretical spectrum.

This calculation is not difficult so much as tedious and if one must carry it out many times per day, use of the computer is indicated. LABDET is an option within

MSSS on the ADP-Cyphernetics network and its use costs $2.00.

N. Conformational Analysis of Molecules in Solution (CAMSEQ). A problem of long standing in chemistry has been to estimate the relationship between the conformation of a molecule in the crystal, as measured by X-ray methods, with that in solution where barriers to rotation are greatly reduced. A sophisticated program set for Conformational Analysis of Molecules in Solution by Empirical and Quantum-mechanical methods (CAMSEQ) has been developed for this purpose by Hopfinger and co-workers [12] at Case Western Reserve University.

This program can run in batch or interactively. As input data, it requires the structure of the compound and this can be provided as a set of coordinate data from X-ray measurements, it can be entered interactively in the form of a connection table or the program can simply be provided with a CAS registry number, and if the corresponding connection table is in the files of the CIS, it will use that.

The first task is to generate the coordinate data corresponding to a particular compound. Then the free energy of this conformation in solution is calculated. Next the program begins to change torsion angles specified by the user in the conformation and with each new conformation, a statistical thermodynamic probability is calculated, based upon potential (steric, electrostatic, and torsional) functions and terms for the free energy associated with hydrogen-bonding molecule-solvent and molecule-dipole interactions.

This program, in its interactive version, can be run in under 20K words of core and CAMSEQ has recently been added to the CIS as part of the X-ray system described in section C of this chapter.

There are a number of other data bases of valuable numeric information which are being built and obtained or expanded from existing sources. These include IR, aquatic toxicity, partition coefficients and thermodynamic data. Anyone interested in the status of these and other such CIS projects is welcomed to write to either of the authors for a copy of the CIS status reports, an informal progress report issued every six months jointly by EPA and NIH.

Conclusions

One of the first goals of the CIS was to produce a series of searchable chemical data bases for use by working analytical chemists with no especial computer expertise. A second aim was to link these data bases together so that the user need not be restricted to a consideration of for example, only mass spectral data.

The various problems inherent in these plans included acquisition of data bases, design of programs, dissemination of the resulting system and linking, via CAS registration numbers, of the various CIS components. These problems, as has been described above, have been solved conceptually and, to a large extent, practically, and the CIS, as it now stands, is the result. It is now possible therefore to review the system in an effort to define future goals, and a number of these seem fairly clear.

Searches through more than one data base in combination would be very desirable. For example, one often possesses mass spectral and nmr data for an unknown and it would be very useful to be able to identify any compounds that match these data in a single search. Work is going on in this area do interface programs so that this approach can be tested. In another development, it is expected that the CONGEN programs developed for the DENDRAL project [30] will be merged into CIS within the next two years. This program, which generates structures corresponding to a specific empirical formula, could be extremely useful in an strategy for structure solving using the CIS. It is not at all difficult to envisage situations in which a reduced set of structures could be produced for considerations by CONGEN. Each structure in turn could be used as an input in the substructure search system and the various compounds whose registry numbers are so retrieved could be considered to be possible answers to the problem.

Confirmation for any of them could then be sought in the spectral data bases, the registry number being all that is necessary to locate and retrieve data. One can even speculate further to the day when synthetic pathways to any likely candidates could be designed by the computer system which could easily add the very practical touch of checking that any starting materials for such syntheses are commercially available at an appropriately low cost!

In a different approach, the power of pattern recognition techniques could be assessed within some of the very large files contained in the CIS. This is a very useful exercise because there is little reported work of this sort on large files and thus we have begun to explore the value of such methods in handling the problem of identification of true unknowns such as water pollutants. Programs designed to test mass spectra for the presence in the compound of oxygen or nitrogen are currently being written [31] and their utility as pre-filters on mass spectral data prior to data base searching will be tested as soon as is possible.

In summary, it is felt that progress to date with the CIS has demonstrated economic feasibility and scientific value in support of TSCA. The test before us is whether we can capitalize on this to explore the new and exciting possibilities that lie ahead in the area of using computer systems, such as the CIS, to support TSCA goals and assist in the demands of the country for a safer and healthier environment.

Literature Cited

1. Heller, S. R., Milne, G. W. A., and Feldmann, R. J., Science, (1977), *195*, 253.

2. Heller, S. R., Fales, H. M., and Milne, G. W. A., Org. Mass Spectrom., (1973), *7*, 107; Heller, S. R., Koniver, D. A., Fales, H. M., and Milne, G. W. A., Anal. Chem., (1974), *46*, 947; Heller, S. R., Feldmann, R. J., Fales, H. M., and Milne, G. W. A., J. Chem. Doc., (1973), *13*, 130; Heller, S. R. and Milne, G. W. A., J. Chem. Info. Comp. Sci., (1976), *16*, 176.

3. Dalrymple, D. L., Wilkins, C. L., Milne, G. W. A., and Heller, S. R., Org. Mag. Res., (1978), in press.

4. Kennard, O., Watson, D. G., and Town, W. G., J. Chem. Doc., (1972), *12*, 14.

5. McCarthy, G. and Johnson, G. G., paper C3 presented as a part of the Proceedings of the American Crystallographic Association meeting, State College, PA., 1974.

6. NIOSH, Registry of Toxic Effects of Chemical Substances, (1976).

7. Unpublished EPA data. For further information one can either Charles Stephan, EPA, Duluth, MN 55804 or Steve Schimmel, EPA, Gulf Breeze, Sabine Island, FL 32561.

8. These include the bibliographic file associated with the data in reference 4 and the Mass Spectrometry Literature Bulletin, published by the Mass Spectrometry Data Centre, Aldermaston, England.

9. Knott, G. D., and Shrager, R. I., Assn. Comp. Machin., SIGGRAPH Notes 6, (1972), 138.

10. Hammer, C. F., Department of Chemistry, Georgetown University, Washington DC 20007, unpublished results.

11. Heller, S. R. and Jacobson, A. E., Anal. Chem., (1972), 44, 2219.

12. Weintraub, H. J. R. and Hopfinger, A. J. Intnl. J. Quant. Chem., (1975), 9, 203; Potenzone R., Cavicchi, E., Weintraub, H. J. R., and Hopfinger, Comp. and Chem., (1977), 1, 187.

13. Gelerntner, H. L., Sanders, A. F., Larsen, D. L., Agarwal, K. K., Boivie, R. H., Spitzer, G. A., and Searleman, J. E., Science, (1977), 197, 1041.

14. Feldmann, R. J., Milne, G. W. A., Heller, S. R., Fein, A., Miller, J. A., and Koch, B., J. Chem. Inf. and Comp. Sci., (1877), 17, 157. and references cited therein.

15. Heller, S. R. and Milne, G. W. A., and Feldmann, R. J., J. Chem. Inf. and Comp. Sci., (1976), 16, 232.

16. EPA Order #2800.2, issued May 27, 1975.

17. The NIH/EPA/MSDC data base is available for lease from the US National Bureau of Standards, Office of Standard Reference Data, A537 Administration Building, Washington, DC 20234. (Telephone 301-921-2467).

18. McLafferty, F. W., Org. Mag. Res., (1978), in press.

19. Clerc, T. L., R. Schwarzenbach, J. Meili, and H. Koenitzer, Org. Magn., Reson., (1976), 8, 11.

20. These data are available as NBS tape #9. Contact the National Technical Information Service (NTIS), Springfield, VA 22151 for details.

21. Hanawalt, J. D., Rinn, H. W., and Frevel, L. K., Ind. Eng. Chem., (1938), 10, 457.

22. Environmental Protection Agency (EPA), Toxic Substances Control Act (TSCA) Inventory reporting Requirements, Federal Register, 42, 247, Friday December 23, 1977, pages 64572-64596. In particular, see section 710.7 on pages 64579-64580.

23. Abramson, F. P., Anal. Chem., (1975), 47, 45.

24. PL 91-596, Occupational Safety and Health Act of 1970 (OSHA), section 20 (a).

25. O'Korn, L. J., Chapter 6 in "Algorithms for Chemical Computations", ed. by R. E. Christoffersen, ACS Symposium Series #46, (1977).

26. PL-94-469, Toxic Substances Control Act of 1976 (TSCA).

27. Vinton, V. A., Milne, G. W. A., and Heller, S. R., Anal. Chim. Acta, (1977), 95, 41.

28. Hartman, K., Lias, S., Ausloss, P. J., and Rosenstock, H. M., Publication NBSIR 76-1061, July 1976.

29. Johannesen, R. B., Ferretti, J. A., and Harris, R. K., J. Magn. Res., (1970), 3, 84.

30. Carhart, R. E., Smith, D. H., Brown, H., and Djerassi, C., J. Amer. Chem. Soc., (1975), 97, 5755.

31. Meisel, W., Jolley, M. and Heller, S. R., in preparation.

RECEIVED November 17, 1978.

INDEX

INDEX